SPACE
CAREERS

SPACE CAREERS

Charles Sheffield and Carol Rosin

Quill

New York   1984

Grateful acknowledgment is made for permission to reprint the following:

Excerpt from *Interplanetary Flight* by Arthur C. Clarke, copyright 1951 by Arthur C. Clarke, published by Harper & Row, reprinted with the permission of the author.

Excerpt from *The High Road* by Ben Bova, copyright © 1981 by Ben Bova, reprinted with permission of Houghton Mifflin.

Library of Congress Catalog Card Number: 84-60210

ISBN: 0-688-03182-X
ISBN: 0-688-03256-7 (pbk)

Printed in the United States of America

2 3 4 5 6 7 8 9 10

BOOK DESIGN BY PATTY   LOW

To Arthur C. Clarke and Wernher von Braun,
whose work, writings, and philosophy
have so influenced the authors
that unwitting plagiarism seems inevitable.

Foreword

Why write a book on space careers?

The best answer to that is through another question, one that has been asked us too many times to count. It comes in a variety of forms, from people of different ages and backgrounds, but usually it goes as follows:

"I have always been interested in the space program and in space exploration. I'm not an engineer or scientist, but I'm willing to work hard and to study. Can you suggest any way that I might find a job connected with space?"

Letters asking the same question arrive each week at the headquarters of the American Astronautical Society, the L-5 Society, the National Space Institute, and similar groups active in promoting space development (see Chapter 4 for an introduction to them). The writers vary widely in age and experience. Some are still in junior high and high school; some are married, have raised families, and at last have enough time to pursue a long-held interest; others have a successful career in another field, but would like to apply their experience and resources to space development.

Although their questions receive sympathetic answers, there is a limit to what can be done for any individual inquiry. We have wished again and again that some general volume existed to which inquirers could be directed. Twenty years ago, two or three good books existed: *Careers in Astronautics and Rockets,* written in 1962 by Carsbie Adams and Wernher von Braun; *Careers and Opportunities in Astronautics,* written in the same year by Lewis Zarem; and *Careers In Space,* written in 1963 by Otto Binder.

All these books were timely and informative. Unfortunately, they now read like museum pieces. In the space age, twenty years have taken us from the first manned flights to the moon, to unmanned exploration of Mercury, Venus, Mars, Jupiter, and Saturn, and to an operating space transportation system built around the space shuttle.

There was another problem, too. In those early years, almost all the job opportunities were seen in engineering and astronautics, and the idea of a service industry for the space program had not yet emerged. Thus, the books of the early 1960's emphasized only *technological* opportunities, and they devoted many pages to general explanations of the space environment. Today, those explanations are unnecessary. Information on space and acceptance of space travel as a fact of life begins before grade school, through books, movies, games, and television programs.

In 1960, earth-resources satellites, communications satellite networks, reconnaissance satellites, and space commerce were thought of as far in the future. Today, space applications are a multibillion-dollar business with good annual growth and tremendous long-term potential. The old books naturally provide nothing to the reader to indicate the jobs provided by space applications.

Reluctantly, we realized that it was pointless referring questioners to the old texts. In the rare cases where they were not long out of print and unavailable, they were too far away from present realities. But there were no newer texts to recommend. The text we wanted, a guide to space careers for the 1980's and 1990's, did not exist.

We finally concluded that there was only one answer: if the book we needed did not exist, it was necessary to create it.

This volume is the result. It divides naturally into three parts. First, we give background on space program activities, and on the general way that we see space program needs developing, first in the near future and then for the rest of this century. Next we address the problem of preparing for a job involving space, including academic training, job applications, and identification of job markets. Finally, we discuss some of the ways to become a "space activist," involved in space affairs even if you cannot at once find work relating to the space program. This includes a discussion of space and politics, of the clouded future of the U.S. space program, and of the emerging international aspects of space development.

Writing a book like this, it is all too easy to find that you are doing little but make lists—of places, addresses, telephone numbers, companies, politicians, budgets, jobs, societies, space organizations, and other related books. This is important and necessary information, but directories make for very hard reading. We have therefore tried to give the essential lists for space career hunting and space involvement, but we have salted this with our own perspectives and perceived priorities for space development, in this country and around the world.

We also want to emphasize that we are looking at a dynamic program. Although we will often sound impatient with the rate of progress in space program development, things are still changing, and changing *fast*. As we write these words the United States has no plan for a permanent manned space station, but a presidential decision and congressional support could change that tomorrow. It's like the Red Queen's race in *Alice in Wonderland*—you must run as fast as you can to stay in the same place. If you want to stay current on space activities, you have to keep reading all the time.

We will try to tell you what to read. But if any reader wants to argue with our opinions, or would like the most recent information on some specific subject, we welcome the dialogue. We can be reached through the publisher, or write to us directly:

Charles Sheffield   Carol Rosin
6812 Wilson Lane   5610 Ten Oaks Road
Bethesda, MD 20817   Clarksville, MD 21029

Acknowledgments

Many organizations and individuals have contributed to the information found in this book. We would like to thank in particular Trudy Bell, Carolyn Brown, Mark Chartrand, David Cummings, Leonard David, Roger Dekok, Dan Deudney, Brian Duff, Lou Friedman, Paul Gardner, Nancy Graham, Richard Gross, Barbara Hubbard, Stan Kent, Howard Kurtz, Igor Makarov, Amy Marsh, Alla Massavich, Jim McGovern, Carolyn Meinel, Ed Mitchell, Jim Muncie, Vladimir Mykoyan, Bill Nixon, John Pike, Kevin Sanders, Wes Thomas, Willard van de Bogart, and David Webb.

All these people helped us with the facts, but we insist on full credit for any errors.

Contents

Foreword   **7**

Acknowledgments   **11**

1 The Road to Space   **17**
The Age of Change   Unpleasant Truths   The High Road

2 Understanding the U.S. Space Program   **26**
The Early Dreams   The Formation of NASA
Putting NASA in Perspective   A Look at NASA Budgets   The Structure of NASA   Outside NASA
NASA Facilities   Recommended Reading About Space, the Space Program, and Its History

3 Training for a Space Career   **56**
A Short Sermon   Multiple Choice   Picking a College   Suggestions for Colleges   Graduate Studies

4 Getting Involved   **76**
Space Organizations and Societies

5 Finding a Job   **118**
Selling the Product   The Changing Job Market
Making a List   Where Do You Look?   Jobs in
Industry   Jobs in Space Command   Retrofitting:
The Problem of a Second Career   State Aviation
and Aerospace Departments   Books on General Career Planning

6 Women in Space   **155**
Why Women?   Slow Progress   Working Rules
Reference Works   Information from NASA

7 Space and the World   **165**
The Canadian Space Program   The European Space
Agency (ESA)   The Japanese Program   The
Swedish Program   The French Program   The
British Program   The Indian Program   The Chinese Program   The Soviet Program

8 Into the Future   **203**
The Long Road   Problems for the U.S. Space
Program   Linkages: Space and the United States
The Military Program   How to Help the U.S. Space
Program   The Hardest Question

9 Index   **235**

SPACE
CAREERS

1

The Road
to Space

"We are at a point in history where a proper attention to
space, and especially near space, may be absolutely crucial
in bringing the world together."

—Margaret Mead

"Nothing endures but change."

—Heraclitus, about 500 B.C.

The Age of Change

This book will not find a job for you in the space program.
It is not intended for that purpose. It is designed to provide
only a guide, a road map that will let you understand your
own strengths and weaknesses, and match those to the general types of jobs available involving space. With a guide in
hand, you will be in a position to do your own job search.
We believe firmly in the old Chinese wisdom: "Give me a

fish and you feed me for a day. Teach me how to fish and you feed me for the rest of my life."

That approach is necessary for several reasons. First, there is no way that we can know every reader's background, age, intelligence, interests, or financial position. All these variables are important. The man or woman who has already raised a family and now wishes to become part of the space program usually has more financial resources than the high school student who is still trying to decide on college courses. On the other hand, companies are less interested in hiring someone who is unlikely to be with them for a long time. Training costs money, and most organizations prefer to train young people. The recognition of simple facts like these can save you a lot of fruitless effort looking for jobs where you will not find them.

Even more important in deciding the structure of this book, we must never lose sight of the fact that we are at the very beginning of the space age. This despite the fact that in thirty years, we have come from a time when "flying to the moon" was used as an expression for something totally ridiculous and impossible to a time where satellite launches are so much a part of routine life that they are noted as news *only when they fail.* Weather satellite pictures are standard on television daily news, and communications satellites carry an increasing proportion of all electronic messages. But even so, space exploration and development remain infants, minute in scope compared with their potential a century from now.

As these fields develop, their needs will change radically. A look at technological changes of the past twenty years should convince anyone that we are on shaky ground if we try to project what will be happening in space in the year 2000, still less for the years beyond. Yet many of our readers will spend most of their working lives after the end of the century.

The one thing that we can be confident of is that there will be new technological breakthroughs. What will they be? We cannot say—one definition of a "breakthrough" is that it is a development that nobody could possibly have predicted. In 1958, when NASA was born, computers were room-filling giants, full of bulky and fickle vacuum tubes. Their software was correspondingly primitive, so that programming was almost a black art. Instruments of equal power, infinitely easier to program, can now be held in the palm of your hand. And the rate of change in the computer field is faster than ever.

In 1958, lasers were years in the future, as were solar-power collectors, "smart" industrial robots, medical telemetry equipment, holograms, computerized body scans, digital watches, and the light but superstrong materials now used widely in aerospace hardware. There will be new engineering gadgets, inventions, and developments at least this striking in the next quarter-century. And even more important in long-term impact may be the biological developments of the third and fourth quarters of the century. The implications of these are not yet clear, but we can look for profound effects from our new understanding of the genetic code, and from new techniques of molecular biology, recombinant DNA manipulations, and gene-splicing. Improbable as it may seem today, these fields will inevitably spill over to affect the course of the traditional engineering sciences and of space-program development. This is a logical certainty, but saying *how* and *when* these things will happen is a different matter.

At the same time, there will be different perspectives on *existing* technology. Techniques will find unexpected applications, just as some methods we now think of as promising will reveal hidden problems (in the 1940's, nuclear-fission plants were described as pollution-free, safe, inexhaustible sources of cheap energy—not quite their image today!).

Taken together, breakthroughs and changes of viewpoint will make great differences to the job opportunities for work on and in space.

So where will the jobs be?

We can't say. All we can do is remain as adaptable as possible, and adjust to the new job markets as they appear. Two hundred years ago, a project to produce a carriage that could travel five hundred miles in a day would probably have spent much of its resources trying to breed a super-horse, one that could run faster and farther than the horses of the day. Few people would have thought the solution might lie in *steam,* a useless and apparently feeble product of boiling water. No one would have thought that burning volatile hydrocarbons in a closed chamber could lead to the modern internal-combustion engine, and an efficient way to produce forward motion.

The other big area of uncertainty is timing. When will we see a mission to Mars, a solar-power satellite, a permanent space station, or unmanned exploration of the surfaces of Uranus and Neptune?

Here, history is encouraging. Estimates for the rate of technological progress have invariably been too conservative, or have projected progress in the wrong fields. Space travel provides some wonderful examples of the problem. Even the most distinguished and intelligent individuals have often had trouble imagining how rapidly technology can advance. So we find, in 1927, J.B.S. Haldane, one of the leading intellects of his generation, predicting the first Mars and Venus landings millions of years in the future.

In 1936, the future astronomer royal of England, reviewing P. E. Cleator's book *Rockets Through Space,* dismissed the whole subject as "essentially impracticable, in spite of the author's insistent appeal to put aside prejudice and to recollect the supposed impossibility of heavier-than-air flight before it was actually accomplished. An analogy such as this may be misleading . . ." Similar criticisms of Cleator's work were still being made in 1953—less than sixteen years before the first manned moon landing!

Finally, we find the noted astronomer Patrick Moore, commenting in 1958 on the possibility of sending a rocket to look at the hidden far side of the moon: "To hope for an early success is being highly over-optimistic." Just the following year, on October 7, 1959, the Soviet spacecraft Luna 3 had flown to the moon, photographed the far side of it, developed the pictures, and transmitted them back to Earth.

The only thing we can absolutely guarantee about future space-program needs is that they will differ from those of today. Thus, the only useful guide to a long-range career involving space begins by accepting rapid change as inevitable, and as a factor that changes career plans.

Unpleasant Truths

About twenty-three hundred years ago, Ptolemy I, the ruler of Egypt, became interested in learning geometry. He arranged for Euclid, the foremost mathematician of the time, to give him lessons; but he found it hard going. What about some shortcuts? he asked. Surely a king should not have to endure hours and days of tedious diagrams, constructions, and chains of logic. Surely there were ways to learn what he wanted, without all those hard and boring proofs.

Euclid's simple answer must have been distasteful to the king: "There is no royal road to geometry."

What can this possibly have to do with finding a career in the space program? It will become clear in the following chapters. Space is certainly here to stay, as an essential and growing part of human affairs. We will never again see a time when no artificial satellites swarm in orbit around the earth; and we can look to the number of jobs that are somehow involved in space steadily increasing, as they have increased for the past quarter of a century.

That sounds wonderful. But we will find that there is no "royal road" to space. Every job we can identify will call for learning, enthusiasm, and hard work—more work than

many jobs in other fields, because the space program is a
rapidly developing, rapidly changing arena. There is a constant need for flexibility of approach, for a willingness to
"think new" about a problem instead of being content with
the old solution.

If you do not like the idea of a continuing challenge, a
space job is not for you. If you love challenge, change, and
hard work, welcome to a career in space.

The High Road

There may be no royal road to a space career, but we can
certainly define a "high road" that leads there. Attend a
good university and major in a technological subject. Graduate near the top of your class, stay in school for graduate
work, and earn a science Ph.D. At the same time, maintain
yourself in tip-top physical shape, with plenty of outside
interests that call for skill, coordination, and nerve. When
you graduate, you will have an excellent chance of moving
to an active role in the space program, and a reasonable
chance of flying in space yourself. Many of NASA's astronauts, men and women, have followed a career path that
looks just like this. It's tough, but possible.

If the course described in the last paragraph is one you
believe you could follow, then you do not need this book.
Your way is already clearly defined. Go and do it, and become an astronaut, or a space scientist.

But there is no need for the rest of us to give up hope. If
we think of the space program as a great pyramid, only a
handful of people up at the very tip will travel to space in
the next ten years. All the rest of the pyramid is made up
of people who lack the extraordinary skills and physical
ability of the astronauts. But the tip of the pyramid cannot
stand high without all the supporting lower layers. The rest
of us play an absolutely essential role in permitting the select few to explore space. In some ways, the particular indi-

viduals who finally fly the manned space program matter less than their support structure of nonastronauts. To quote Jesco von Puttkamer of NASA, "NASA succeeded as spectacularly as it did in large measure by impersonalizing space flight, by replacing single efforts and more limited crusades with the staggering power of regimented attack by four hundred thousand well-organized and highly motivated people in government, industry, and universities, backed up by strong congressional support and the commitment of a President."

The jobs in those "supporting layers" may sound much less romantic than piloting a spacecraft; but they greatly increase the number of types of people who can play an active part in space-program development. Compare the number of people who build, service, schedule, and arrange flights on commercial aircraft with the few who pilot them. The support jobs for space are in medicine and health, law, technical writing, engineering, computing, communications, marketing (space is going to be a very big business), public relations, psychology, and education. Although when we think of space we think of NASA, *most of the future jobs will not be with the government.* They will be industrial positions, and as space activities develop they will be available in the thousands and the hundreds of thousands. Even at the peak of the Apollo Program, far more people were working for industrial contractors than for NASA itself. And that is still true today—the space shuttle is not built by NASA employees, it is built by the employees of U.S. companies. In fact, almost all the spacecraft and equipment involved in the U.S. space program have been built on contract. NASA's job today is overall program guidance and management.

"Program guidance and management"; that may not sound too exciting to most of us. We are a long way, you might say, from the dream of flying on board a spacecraft. But not really. We must also remember that space travel will become less demanding, physically and mentally, as time goes on. Opportunities to go into space will increase, and become commonplace.

As a useful analogy, we can look back to the beginning of this century. At that time, air travel was something reserved for daredevils of exceptional endurance and nerve. And now? Eighty-year-olds think nothing of intercontinental travel. By the close of this century, space will be open to men and women who have no exceptional qualities—except the strong desire to go there.

When those opportunities come along, they will come most often and earliest to those who have had the foresight and self-discipline to prepare themselves. We should anticipate no shortage of well-prepared individuals, so the initial competition will be strong.

"Produce ships and sails that can be used in the celestial atmosphere," said Johannes Kepler, in a letter to Galileo written at the beginning of the seventeenth century. "Then you will also find men to man them, men not afraid of the vast emptiness of space."

That is certainly true today. But we would add, "and women too." As many women as men seem to be interested in space careers; more and more jobs are seen as performed equally well by both sexes.

The larger task is the growth of the space program itself. We will suggest ways that you can help with this in more detail in Chapters 4 and 8, but let us point out here that a long training is not necessary before your involvement can begin.

A strong, weapons free, creative space program stems from the support and wishes of many people, spread through the whole country, and the world. No matter how strongly you feel about space, no single individual can create that national desire. We can all contribute to it, and encourage the nation's and world leaders to listen to our views. Ultimately, the U.S. elected leaders are the ones who will vote for more or less in the United States space program. But they listen very closely to a vocal electorate. For as Jacques Cousteau remarks: "Explorations have to be inspired and triggered by a leader, but they have to meet with the acceptance and the enthusiasm of the crowd."

2

Understanding
the U.S. Space Program

"If I could get one message to you it would be this: The
future of this country and the welfare of the free world depends upon our success in space. There is no room in this
country for any but a fully cooperative, urgently motivated
all-out effort toward space leadership. No one person, no
one company, no one government agency, has a monopoly
on the competence, the missions, or the requirements for
the space program."

—Lyndon Baines Johnson

To many people, the center of the United States space program may appear to be at the Johnson Space Center in
Houston; or perhaps it seems to be in Florida at the Kennedy Space Center, the site for the most spectacular NASA
launches.

Those ideas are quite wrong. The center of the U.S.
space program today is where it has always been—in Washington, D.C. The situation could change by the end of the

century, but for the moment any real understanding of the space program must begin in the nation's capital. And the story should begin a long time ago, back in the days when even air travel was considered dangerous, unusual, and unnecessary.

If we want to be part of the United States space program, we should know its history, how it came into being, what its operating rules are, and how it fits into the space activities of the rest of the world. In this chapter, we have space for no more than a brief summary of a long and complicated process of development. At the end of the chapter, however, we list a number of books that explore the subjects of the United States and world space programs, past, present, and future, in great detail.

Our list does not seek to be exhaustive—in the past quarter of a century hundreds of books have been written about space. In making the selection, we have chosen books of unusual accuracy and readability. Also, since many good books about space are unfortunately out of print, we have generally preferred more recent and readily available texts over older or less accessible ones.

The Early Dreams

Humankind must have dreamed of flying like a bird for many thousands of years. That desire certainly came before there was written history—perhaps before there was development of language. The old legend of Daedalus and his son, Icarus, who flew too near to the sun, is evidence that the longing to fly was well established thousands of years ago. But for most of history, the dream of flying had to be just that, an impossible dream.

Fulfilling the dream came closer in the late eighteenth century, when in 1783 the Montgolfier brothers built the first large hot-air balloon. In the rest of the century, manned balloon flight became accepted, even if not exactly common. But such flights were completely at the mercy of the winds. Directed flight in a heavier-than-air machine seemed unattainable, and learned papers were written through the nineteenth and early twentieth century, proving that it was impossible or would never be important (birds were presumably regarded as an optical illusion, or somehow as an example that didn't count).

> "Heavier-than-air flying machines are impossible."
> —Lord Kelvin, 1892

> "The demonstration that no possible combination of known substances, known forms of machinery and known forms of force, can be united in a practical machine by which men shall fly long distances through the air, seems to the writer as complete as it is possible for the demonstration to be."
> —Simon Newcomb, 1900

> "Flight by machines heavier than air is unpractical and insignificant, if not utterly impossible."
> —Simon Newcomb, 1902
> (This last statement was made
> only a year and a half before
> the Wright brothers' first flight.)

> "The aeroplane will never fly."
> —Lord Haldane (British minister of war), 1907
> (This statement was made four years
> *after* the Wright brothers' first flight!)

Progress toward heavier-than-air flying machines took a tremendous stride forward in the late nineteenth century, when the internal-combustion engine was developed. It became a reality in 1903, when Orville and Wilbur Wright flew the first airplane at Kitty Hawk, North Carolina.

But by the time that humankind gained access to the ocean of air above us, other developments suggested that the moon, planets, and stars were even more inaccessible than they had seemed to the ancients. Distances were immense, hundreds of millions of miles to the planets, trillions to the stars. Flight now appeared possible only where there was air to provide lift. Since Earth's atmosphere dwindles away to imperceptible traces less than thirty miles above the surface, there is nothing beyond that against which a propeller could push, nothing from which a balloon could derive lift. Humankind appeared to be firmly bound within the thin envelope of air that surrounds the world.

By the time that human flight was achieved, man had already learned that the other planets of the solar system were whole worlds, some of them tantalizingly similar to the earth in size and composition. The nearest and easiest target naturally appeared to be the moon. For hundreds of years, long before people understood the true nature of the difficulties of space travel, humans had studied our nearest neighbor in space and proposed fanciful ways to get there. The methods ranged from Cyrano de Bergerac's bottles of dew, drawn into the air by the sun's heat, to Francis Godwin's flights of migrating birds.

By comparison with these, Jules Verne's idea of a giant cannon, built by the Baltimore Gun Club, set deep into the soil of Florida, and capable of firing a projectile all the way to the moon (in his 1865 novel, *From the Earth to the Moon*) looks like hard-headed and practical science. Unfortunately, it, too, would not work. The forces experienced by its passengers would make them weigh twenty-two thousand times as much during the launch as they did on Earth. By the time that they reached the end of the gun barrel, the brave travelers would have been reduced to a thin smear of protoplasm in the bottom of their space capsule.

The final realization that there *is* a way to go into space, and a practical and feasible way at that, was made by three men, one Russian, one American, and one German. The present space program, of every country, owes a great debt of gratitude to Konstantin Tsiolkovski, to Robert Goddard, and to Hermann Oberth. Their work was not well appreciated at the time, even by people who should have known better. One reaction to Goddard's early efforts appeared in a 1920 *New York Times* editorial. (*The Times* finally printed a correction to this editorial comment—fifty years later.):

". . . after the rocket quits our air and starts on its longer journey, its flight would be neither accelerated nor maintained by the explosion of the charges it then might have left. To claim that it would be is to deny a fundamental law of dynamics, and only Dr. Einstein and his chosen dozen, so few and fit, are licensed to do that.

"That Professor Goddard, with his chair in Clark College and the countenancing of the Smithsonian Institution, does not know the relation of action to reaction, and of the need to have something better than a vacuum against which to react—to say that would be absurd. Of course he only seems to lack the knowledge ladled out daily in high schools."

Tsiolkovski, Goddard, and Oberth all saw that the key to space travel was the rocket, which can use reaction to its own exhaust gases to permit it to travel in any direction, even in a vacuum. Tsiolkovski set forth the basic principles for the use of rocket propulsion in his 1903 work *Investigation of Cosmic Spaces by Reactive Devices*. However, he was a theoretician, not an experimenter. He foresaw many of the key developments of today's space program—he had early in this century pointed out the value of liquid hydrogen as a fuel (in 1903, only five years after hydrogen had first been cooled to a liquid state!). But he did not try to build the rockets he described. It was Robert Goddard who turned theory to practice, and in 1926 launched the first liquid-fueled rocket.

And the public response? As that *New York Times* editorial suggests, it ranged from mild interest to misunderstanding and ridicule, even by people who should have known better:

"This foolish idea of shooting at the moon is an example of the absurd length to which vicious specialization will carry scientists working in thought-tight compartments. Let us critically examine the proposal. For a projectile entirely to escape the gravitation of the earth, it needs a velocity of 7 miles a second. The thermal energy of a gramme at this speed is 15,180 calories . . . The energy of our most violent explosive—nitroglycerine—is less than 1,500 calories per gramme. Consequently, even had the explosive nothing to carry, it has only one-tenth the energy necessary to escape the earth . . . Hence the proposition appears to be basically impossible." (A. W. Bickerton, 1926. Professor Bickerton had problems; not only could he not conceive of a staged rocket, but he appeared to confuse the desire to launch the payload with a desire to launch the fuel. Also, nitroglycerine, an excellent explosive, has far less available energy per gram than dozens of other fuels.)

Goddard was aware of and sensitive to these criticisms, but fortunately he did not let them affect his actions. In 1898, when still a youth of sixteen, he had set his sights on finding methods to travel beyond the earth's atmosphere. He continued to pursue them steadfastly until his death in 1945. As he said in a 1932 letter to H. G. Wells, "How many more years I shall be able to work on the problem, I do not know; I hope, for as long as I live. There can be no thought of finishing, for 'aiming at the stars,' both literally and figuratively, is a problem to occupy generations, so that no matter how much progress one makes, there is always the thrill of just beginning."

The Formation of NASA

Goddard's achievement was not appreciated in this country, except by a few people such as Charles Lindbergh. It *was* appreciated, however, in Germany. During the Second World War, a liquid-fueled rocket using Goddard's ideas became the basis for a weapon. This was the V-2, a ballistic rocket that could deliver a one-ton explosive warhead to targets several hundred miles away. It reached them by flying out beyond the earth's atmosphere. When the United

States announced in 1955 that it would launch a small, unmanned earth satellite as part of the International Geo-
physical Year in 1958, the basic rocket design drew heavily
on that of the V-2. However, while America was still planning, the Russians, in October 1957, launched the first artificial earth satellite, Sputnik 1—with a rocket whose
design also drew much from the V-2.

The United States, until this time confident of its technological lead over the rest of the world, was shocked. It
soon became doubly shocked when the first U.S. attempt to
launch a small satellite, no bigger than a grapefruit, ended
as a disastrous failure (with live television coverage, to rub
salt in the wound).

The government responded rapidly. By 1958, the National Aeronautics and Space Administration had been
formed from the old NACA, the National Advisory Council for Aeronautics, and the U.S. space program was on its
way.

It is important to understand NASA's origin. The agency
was brought into existence as an alarmed response to a perceived Russian technological superiority during the cold
war. It would make use of a technology that had been developed mainly as a weapon in the closing stages of a world
war. And its most famous single achievement, the landing
of humans on the moon, would be a feat whose motivations
were political and ideological. Given that background, and
the tense international situation of the day, it is almost mi-
raculous that the Space Act of 1958 could contain this important statement of intent:

*"It is the policy of the United States that activities in space
should be devoted to peaceful purposes for the benefit of all
mankind."*

This set the stage for the remarkable series of NASA
missions that we have seen over the past quarter of a century. And most of them we saw on television—a possibility
that no fiction writer ever prophesied in describing the
United States' move into space.

NASA was established with an "open skies" policy. If there were to be failures, they would be public failures. Reporters and camerapeople can be present, if they choose, at every launch. They have access to all the technical details, of design, operations, and occasionally of malfunctions.

When an exploding oxygen tank on Apollo 13 damaged the command capsule's electricity, computer, fuel cells, and water supply (on April 13, 1970—food for thought for the superstitious), the whole world was watching and listening. Hour by hour reports detailed the brilliant series of improvised actions that took the crew of three astronauts around the moon, and finally back safe to Earth.

With this public mode of operation for NASA, it ought to be clear that although the agency's activities are an important and highly visible part of the United States space program, they are not the whole story. The existence of special surveillance satellites, with cameras that can take pictures of the earth's surface with extraordinary detail, and every launch on the planet, is a poorly kept government secret. There have been articles about them in many popular magazines—sometimes giving details of their manufacture and performance.

There is thus a substantial military space program, which does not operate under an open-skies doctrine, and whose size and scope is little publicized. This **is** also a potential area for space-related careers, as is the growing nongovernment space activity. In looking at the total career opportunity, it is thus essential to know about the activities beyond the NASA programs. But first let us look at NASA in a little more detail.

Putting NASA in Perspective

Whenever we read of U.S. space activities, NASA is usually involved. There is a tendency to think of it as a vast and powerful agency, well funded in comparison with other branches of the government. That idea is unfortunately far from the truth. NASA is minuscule compared with the Defense Department, or the Health and Human Services Department, or many other government activities.

Let's look at some numbers.

The NASA budget for 1983 was $6.8 billion. To most of us, that is a huge figure—even an incomprehensibly large figure. But we should compare it with the total Defense Department budget proposed for 1984. That is $245 billion, not including the costs of veterans' pensions and benefits. The United States spends twenty-eight million dollars on defense, every hour of every day, compared with much less than a million on space. The NASA allocation for an entire year would run the Defense Department for just eleven days.

The Health and Human Services proposed budget is even larger, at $373 billion. From a total proposed federal budget of $848 billion for 1984, NASA will receive less than one percent. Reducing our space spending—even reducing it to zero—will not eliminate the federal deficit. It will not even make a noticeable dent in it.

It is also easy to forget that one of the A's in "NASA" stands for *aeronautics*. NASA's budget must support not only space, but this country's major aeronautical research. Since aircraft and air equipment are this country's second biggest export, exceeded only by agricultural products, this investment on research has carried an excellent return.

It is also surprising to note that the Defense Department spends more money per year *on space* than the whole NASA budget. This has been true for at least the past four years. The whole of the U.S. space program is thus clearly not a civilian activity. It is not even predominantly so, and many space-related job opportunities exist as part of military programs and projects.

A second common misconception concerns the size of the NASA budget in absolute terms. Suppose that we think of NASA as a corporation, and measure its revenue against that of other private companies. Where would it stand relative to the major industrial corporations of this country?

The answer may again be a surprise. NASA would not be in the top thirty. Ten oil companies have larger annual revenues than NASA's operating budget. So do many other familiar names such as IBM, Procter and Gamble, Dow Chemical, Western Electric, Ford, Chrysler, Westinghouse, RCA, General Electric, and Eastman Kodak. The largest U.S. companies, General Motors and EXXON, have ten times the annual revenue of NASA. And this does not even consider the nonindustrial giants, such as the insurance companies and the large banks, whose revenues also are far in excess of NASA's. We spend many times as much on travel, on cigarettes, and on alcohol as we do on space.

Part of the reason for NASA's small size compared with many other government agencies and private companies arises from the basic NASA charter. It is an agency for *research and development,* not for construction and operations. Once a space project (such as communications satellites, weather satellites, and earth-resources satellites) has become far enough developed that it can be run on a routine, operational basis, NASA will turn it over to some other group. In the examples quoted, the weather and earth-resources satellite programs are now run by the National Oceanic and Atmospheric Administration (NOAA), which is part of the U.S. Commerce Department, and they may soon be turned over to industry operation; and the commercial communications satellite program is already run by Comsat, RCA, and other industry groups. In each case, NASA is still responsible for research activities, but these are usually a small part of the total operation, at least from the point of view of the budget (and jobs) assigned to them.

There is one major exception to the situation stated above. NASA still serves as the agency responsible for all U.S. civilian satellite launches, even though this is very much an operational rather than a research function. It is likely that this, too, may change in the next ten years, as industrial groups become more active in the development of commercial launch capability.

One other point is relevant to anyone seeking a job with NASA. Although the agency is program manager for numerous space-related activities, most of the construction, computer programming, and operations are carried out not with NASA employees, but with contractor personnel. At the most intense stage of the Apollo Program, the NASA direct government payroll had thirty-six thousand people— a big increase from its 1958 strength of slightly less than eight thousand. However, the number of people working on NASA's programs, including contractor employees, was close to four hundred thousand—with three out of four directly engaged on the manned-spaceflight effort leading to the Apollo landings. Less than 10 percent of the total work force were employed in government positions.

The situation has changed little today. There are many more jobs involving space to be found *outside* the government than inside it. However, it is still true that a large fraction of those industry jobs are supported by government funds.

A Look at NASA Budgets

The amount of money appropriated to NASA varies, year by year. This amount is decided by a complex interaction between the White House, the Office of Management and Budget (OMB), NASA, and the Congress (the whole process is outlined in Chapter 8). Every major new project forms a "new start," which must have funds approved before work can begin. There is no way that NASA could decide to go ahead with a space station, or a solar-power satellite, to choose just two substantial examples, unless these other groups agreed. That agreement is a complex and a time-consuming process. Politics is involved, and so are the international situation, the overall government budget deficit, and the needs of ongoing NASA programs. For example, in the period 1979–1983 the need to continue space shuttle development dominated NASA's budget, at the expense of many other programs.

Table 2.1 shows the NASA budget for the period 1959 to 1983. The same information is presented in graphic form in Figure 2.1. The graph seems to make the overall trend of funding quite clear. NASA had more to spend in 1983 than it has ever had before, including the peak years of the Apollo Program. In fact, Table 2.1 is quite misleading, because it takes no account of inflation. Thanks to inflation, a dollar of budget next year goes less far in its buying power than the same dollar a year ago—and far less than that dollar ten years ago.

No two sources will quite agree on what the inflation rate is in any year (it very much depends on what you are buying—computers can go down, while fuel goes up). But making use of one of the most widely accepted general inflation indices, the Consumer Price Index, the adjusted NASA budget for the period 1959 to 1983 is shown in Table 2.2, all in terms of the buying power of 1959 dollars.

Now the situation looks quite different. In terms of what it can buy with its budget, NASA is far below its peak year of 1965, and has been at a fairly steady or declining funding level ever since. For the past ten years, the budget has held almost constant, at about two billion 1959 dollars. The other thing to be remembered in looking at NASA's budget is that throughout the agency's history there has always been some dominant effort that took most of the funding. At first it was the early attempts to orbit a satellite, then to orbit a man. Through the 1960's, the big project was the Apollo Program, targeted at placing a man on the moon by 1970, and in the seventies the major efforts were Skylab and the development of the space shuttle. Today, the largest item in NASA's budget remains the shuttle, but now it has moved from a development effort to its use as a practical space-transportation system.

Table 2.1

NASA budget, in millions of dollars.

1959	331
1960	523
1961	964
1962	1,825
1963	3,673
1964	5,100
1965	5,250
1966	5,175
1967	4,966
1968	4,587
1969	3,991
1970	3,746
1971	3,311
1972	3,307
1973	3,406
1974	3,037
1975	3,229
1976	3,550
1977	3,818
1978	4,060
1979	4,549
1980	5,240
1981	5,518
1982	5,936
1983	6,836

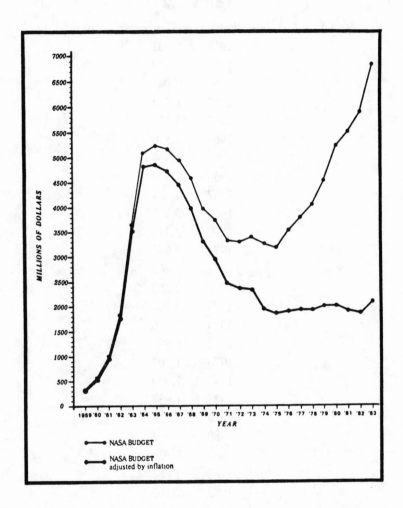

Table 2.2

NASA budget, in millions of dollars, adjusted for inflation.

1959	331
1960	519
1961	941
1962	1,763
1963	3,510
1964	4,816
1965	4,894
1966	4,742
1967	4,424
1968	3,972
1969	3,317
1970	2,954
1971	2,465
1972	2,360
1973	2,354
1974	1,976
1975	1,893
1976	1,907
1977	1,939
1978	1,937
1979	2,017
1980	2,087
1981	1,936
1982	1,887
1983	2,103

It goes without saying that most of the jobs, either in NASA or in the aerospace industries that support it, will be found in those major projects that consume most of the funds. If the biggest effort of the 1980's, as now seems probable, is toward a space station, we can look for continued opportunities in vehicle development. There will also be a new emphasis on three other areas: medical support for long-term living in space; materials and techniques for large space structures; and experiments in space manufacturing. All these areas are likely to become important during the 1980's.

The Structure of NASA

Although NASA forms a single agency of the U.S. government, for many purposes it is better to think of it as a dozen separate pieces. There is a headquarters operation in Washington, D.C., which is responsible for interactions with other government agencies and with Congress, and which serves as the general coordinator of all programs in space and aeronautics research; and there are eleven field centers, where the technical work of NASA is performed.

These field centers all report to NASA headquarters, but they have a good deal of independence, and in many ways compete with each other for funds and interesting programs. They are scattered around the United States.

1. *Ames Research Center* is at Moffett Field, California, not far from Palo Alto. It specializes in stability and guidance of aircraft and spacecraft, in space environmental physics, in simulation methods, and in biomedical and biophysical research.
2. *Goddard Space Flight Center* is in Greenbelt, Maryland, not far from Washington, D.C. It is a primary site for development of unmanned satellites, for astronomical research, earth resources, weather, and com-

munications. This center also operates the NASA tracking and data-acquisition networks.

3. *Dryden Flight Research Center,* part of Ames Research Center, is at Edwards Base, California. It tests high-performance aircraft and spacecraft, as well as general aviation models, and performs research on structures of aeronautical vehicles. The base is also the principal landing site for the space shuttle.

4. *Marshall Space Flight Center* is in Huntsville, Alabama. During the Apollo Program, this center built the large Saturn rockets that formed the first stage of the Apollo missions. Today, it concentrates on launch vehicles and systems for both manned and unmanned space flights, and also integrates payloads and flight activities. About half of Marshall's efforts are in support of the shuttle program.

5. *Kennedy Space Center* is on Merritt Island, on the east coast of Florida. It is this country's primary launch facility, and has been recognized as such since the first days of the U.S. space program. The center prepares and integrates space vehicles. It was the launch site for Mercury, Gemini, Apollo, and Skylab flights, and it is the primary launch and recovery base for the space shuttle. (However, it is not used for the launch of satellites that will orbit the earth's poles, since their upward flight from Kennedy would take them over inhabited areas. Such launches are made from the Western Test Range at Vandenberg Air Force Base, in California.)

6. *Langley Research Center* is located in Hampton, Virginia, just north of Norfolk. It was an original center for aeronautics research, in the days of the National Advisory Committee for Aeronautics. Today, it continues as a prime site for aeronautics research, with extensive computer facilities and high-speed wind tunnels for aerodynamic simulation and testing. It is also a leader for the development of space structures and materials.

7. *Johnson Space Center* is located just outside Houston, Texas. It is the center for astronaut training, and for the development of manned spacecraft and life-support capabilities. This center plans, builds, and integrates experiments for space-flight activities and applications of space technology, and also serves as the lead facility for the space shuttle development and operations.

8. *Lewis Research Center* is in Cleveland, Ohio. It is concerned with development of advanced aircraft engines, and with power plants and propulsion systems for spacecraft. This center also works on engine noise and pollution reduction.

9. *Wallops Island Flight Center,* part of Goddard Space Flight Center, is located on the Virginia portion of the Delmarva Peninsula. It performs launches of small suborbital and orbital missions and atmospheric probes, and also provides tracking services.

10. *The National Space Technology Laboratories* at Bay St. Louis, Mississippi, is engaged in the test-firing of large space and launch vehicles' engines, and supports the main engine and main orbiter propulsion system for the space shuttle.

11. *The Jet Propulsion Laboratory,* located at Pasadena, California, is thought of by many people as NASA's center of excellence for solar-system exploration. The high reputation is well deserved, but in fact JPL is not truly a NASA center. It is part of the California Institute of Technology, and operates under contract to NASA. JPL manages the Deep Space Network, handles lunar and interplanetary data acquisition and data reduction, and formulates methods for orbit computation and analysis.

In addition to these field centers, there are other offshoot facilities associated with them. These include the Michoud Assembly Facility in Louisiana, which is associated with the Marshall Space Flight Center and is responsible for the space shuttle external tank; the Goddard Institute for Space Studies in New York, which supports theoretical research in space science and applications; the Johnson Space Center's White Sands Test Facility at Las Cruces, New Mexico, which supports the space-shuttle propulsion system, power system, and materials testing; and the Western Test Range of the Kennedy Space Center, already referred to as the launch location for polar-orbiting satellites.

The one-paragraph descriptions given here can provide no true idea of the variety of applications tackled at each of these centers. All of them support education programs, electronics programs, and need a wide variety of general support skills, from manning the reception desks to running the cafeterias. To get an idea of the specific hiring policies and needs of each center, write to their public-information offices and their personnel offices. The addresses are given at the end of this chapter.

Outside NASA

The space-related activities of this country can be thought of in four major areas (non-U.S. activities are the subject of Chapter 7). We have:

NASA programs
Civilian government programs other than NASA
Defense programs
Industry programs
(Most states also have an aerospace and aviation department, but they usually focus completely on aviation. Their addresses can be found in Chapter 5.)

The discussion of budgets earlier in this chapter makes it clear that many space-related jobs in the United States will come from military programs; a closer look at defense budgets and space-career potential is thus appropriate.

Defense job opportunities: The total NASA budget for space activities (i.e., excluding aeronautics) and the Defense Department budget for space for the period 1980–1983 are as follows:

(In millions of dollars)

	NASA	Defense
1980	4,700	3,850
1981	5,000	4,800
1982	5,500	6,400
1983	6,100	8,500

(Note: There are probably a couple of billion more space-related expenditures in the 1983 defense budget, which are not explicitly tagged for space).

The rapid increase in defense space spending reflects two factors. First, in the past few years there has been an increased worry about this country's military position, and a corresponding rapid increase in all types of defense expenditures. However, the increases in spending actually began before this concern. It happened because in the past few years there has been a major change in the Defense Department's general attitude to space.

Just a few years ago, most of this country's upper-level military personnel regarded space activities as peripheral to the country's main defense systems. The enthusiasm for space, if it could be found anywhere in the Pentagon, was in the lower ranks. "Nobody with more than two stars has the least interest in space" seemed to be the common wisdom. That has changed dramatically. Space is now seen by the Defense Department as a definite arena for increased unmanned military activity, and perhaps for a larger manned program also. As a result, new activities, many in the area of career development and recruiting, have come into existence in the Defense Department's space program.

Those developments, and what they mean in practical job-hunting, will be discussed in detail in Chapter 5. We will also provide there a review of the types of openings available. Here we will only point out three things. First, military space activities provide numerous *civilian* job opportunities. Second, if you are of the right age and interests, military positions can often provide funding for a college education, since one recognized shortage in the military space programs is for technically trained people. Finally, the biggest near-term increase in space careers seems certain to involve the military programs—let us hope of a new, nonweapons nature.

Space activities of other government agencies: Taken together, NASA and the Defense Department enjoy the lion's share of government space funds. We can summarize all the other space expenditures most easily by showing their percentage of total U.S. government space budgets. The distribution (for 1983) looks as follows:

Agency	Percentage	Comment
NASA	41.17	
Defense	57.37	
Commerce	1.01	Mainly weather-satellite program.
Energy	0.21	
Agriculture	0.16	Mainly for crop-assessment experiments from space.
Interior	0.06	For operation of EROS Data Center and provision of earth resources data.

When we note that the government is considering private ownership and operation of the weather and earth-resources satellites, and that the crop-assessment experiments of the Department of Agriculture are coming to an end, the overall message is clear: Government jobs related to space, other than those supported by NASA or the Defense Department, are few in number. Those two agencies together have more than 98 percent of this country's space funding.

Jobs in private enterprise: There are few space-related

jobs to be found in state and local government (it is tempting to say there are none, but we can't prove that). This leaves us with industry programs still to be considered. A look at the "Help Wanted" section of the newspapers seems promising, since there are many openings for aerospace positions. However, that is misleading. Most of the jobs are with companies who are performing contracts for NASA and the Defense Department, so that money has already been taken into account in discussing the federal space budgets.

What about the truly commercial space job openings, not derived from government contracts?

The position here is not very encouraging, but should change soon as more appreciation of the potential marketplace emerges (especially as a more desirable alternative to a space arms race). Private enterprise in space is still in its early infancy. We can recognize three main areas of development:

- a communications satellite industry, already fairly well developed
- a healthy interest in space manufacturing, being pursued by joint industry-government efforts with McDonnell Douglas and a number of other companies, interesting but still far from the commercial state
- a private launch activity, very much a fledgling enterprise at this point, now being established by small groups of entrepreneurs

The best-known group in communications-satellite work in this country is probably Comsat (Communications Satellite Corporation), a legislatively chartered corporation established by Congress in 1962. Comsat serves as the U.S. representative on Intelsat, the organization that handles worldwide message transfer via satellites. Many other corporations in this country are involved with communications satellites, either as builders (for example, RCA and Hughes) or as operators (for example, Western Union).

Satellite Business Systems was formed by Comsat, IBM, and Aetna Life Insurance, to exploit the use of communications satellites for high-volume data transfer.

Job opportunities with all these groups will be dealt with in Chapter 5. However, we note here that of *all* space activities, this is the only one that today has reached the status of a profitable, mature industry. Space communications, like all communications, is growing at a great rate, with no end in sight. If recent projections for the markets in direct broadcasting satellites and teleconferencing are anywhere near correct, the growth in space communications will be more than rapid; it will be explosive.

By contrast, space manufacturing is very much in the early experimental stages. We are hindered considerably in the United States by the lack of a space station, in which long-term manufacturing experiments can be conducted by human crews. The short flights of the space shuttle, only a week or two, permit only the simplest and most short-lived procedures. Few U.S. jobs can therefore be expected in this area until this country has a long-duration facility for experimentation, similar to that provided by the Russian Salyut 6 and Salyut 7 stations. Even more jobs will become available as we begin to work together internationally, creating new formulas for space development based on global perspectives.

Companies looking into commercial manufacturing potential in Earth orbit include:

- Fairchild Space and Electronics of Germantown, Maryland, who are looking at a for-rent satellite known as Leasecraft
- Ball Aerospace Systems, of Boulder, Colorado, planning another for-rent materials processing facility
- Space Industries, Inc., of Houston, Texas, who are looking at the commercial potential of a manned space station
- McDonnell Douglas Astronautics of Long Beach, Califor-

nia, who have already flown an electrophoresis experiment on the fourth shuttle mission

To encourage this sort of effort, NASA now permits special joint ventures between government and industry. In these ventures, the shuttle is used free of charge as the vehicle on which microgravity experiments are conducted and no government money is paid to the company for its work, but the company is allowed to retain patent rights from the manufacturing process. This area is steadily developing, but the future jobs potential cannot easily be guessed.

Finally, the private launch activity is coming to a most interesting state. In 1982, Space Services, Inc., of America (SSI), a U.S. group funded without a penny of government backing, performed its first suborbital launch of a solid fueled rocket, the Conestoga I, from Matagorda Island off the coast of Texas. Their plans next call for a first orbital launch in 1983 or early 1984, and for regular launch of commercial payloads by 1985. The success of the European launcher, Ariane, in winning contracts around the world has pointed to a large market for inexpensive and reliable launch vehicles. SSI hopes to capture a substantial fraction of that market.

And for job opportunities? It's too early to say, but there is a chance that in the next couple of years private industry (SSI or some other group) will be able to take over all the launches using expendable rockets (i.e., excluding the space-shuttle launches) now handled by NASA. And another group, the Space Transportation Company, Inc. (SpaceTran), has been exploring the possibility of buying a fifth shuttle orbiter, once intended for inclusion in NASA's shuttle fleet. NASA budget restrictions prevented that, but the number of launches required through the 1980's suggests that the fifth orbiter could be kept fully busy. SpaceTran has attracted some substantial backers, including Prudential Insurance Company.

The third component in commercial launch operators is provided by Arianespace, a European (largely French) private group organized to market the services of the Ariane launcher. This is described in Chapter 7 when we discuss the French space program.

Together, these private launch groups add up to a tantalizing picture: lots of interest, lots of promise, and in a few years' time, perhaps lots of jobs. Today, however, they employ no more than a few dozen people. Your best job opportunities still lie with the aerospace and other companies that support NASA and the Defense Department's space work. We want to say it again for emphasis: *Ninety-five percent of all today's space-related jobs are with aerospace companies and support contractors, and they are funded by government programs.*

NASA Facilities

Job inquiries should be addressed to the Personnel Office, general inquiries to the Public Information Office. Each field center also operates an Education Office, from which films and other educational materials are available for loan (usually for free). The states served by each Education Office are listed following the names and addresses of the centers.

NASA Headquarters
400 Maryland Avenue
Washington, D.C. 20546

Ames Research Center
Moffett Field, California 94035

Goddard Space Flight Center
Greenbelt, Maryland 20771

Dryden Flight Research Center
Edwards, California 93523

Marshall Space Flight Center
Alabama 35812

Kennedy Space Center
Florida 32899

Langley Research Center
Hampton, Virginia 23365

Johnson Space Center
Houston, Texas 77058

Lewis Research Center
21000 Brookpark Road
Cleveland, Ohio 44135

Wallops Flight Center
Wallops Island, Virginia 23337

National Space
Technology Laboratories
NSTL Station, Mississippi
39520

Jet Propulsion Laboratory
4800 Oak Grove Drive
Pasadena, California 91103

Certain NASA field centers are set up to answer inquiries from specific states in their education offices. The matching centers and states are as follows:

Ames Research Center: Alaska, Arizona, California, Hawaii, Idaho, Montana, Nevada, Oregon, Utah, Washington, Wyoming.

Goddard Space Flight Center: Connecticut, Delaware, District of Columbia, Maine, Maryland, Massachusetts, New Hampshire, New Jersey, New York, Pennsylvania, Rhode Island, Vermont. (Note: for film loans if you live in Connecticut, Maine, Massachusetts, New Hampshire, New York, Rhode Island, or Vermont, instead of Goddard Space Flight Center, you should contact the National Audiovisual Center, General Services Administration, Washington, D.C. 20409.)

Marshall Space Flight Center: Alabama, Arkansas, Iowa, Louisiana, Mississippi, Missouri, Tennessee.

Kennedy Space Center: Florida, Georgia, Puerto Rico, Virgin Islands.

Langley Research Center: Kentucky, North Carolina, South Carolina, Virginia, West Virginia.

Johnson Space Center: Colorado, Kansas, Nebraska, New Mexico, North Dakota, Oklahoma, South Dakota, Texas.

Lewis Research Center: Illinois, Indiana, Michigan, Minnesota, Ohio, Wisconsin.

Recommended Reading About Space, the Space Program, and Its History

The books that follow are listed alphabetically by author. We regard all of them as well produced, excellent reading, and filled with valuable information and insights. In many cases, these works have appeared in both hardcover and paperback forms. Where this is the case, the year of publication generally refers to the paperback edition. We have tried to list only books that are still in print and can be ordered either directly from the publisher or through any sizable bookseller. In the case of volumes produced by professional societies, consult Chapter 4 for the relevant mailing address.

Bound for the Stars, by Saul and Benjamin Adelman. Englewood Cliffs, NJ: Prentice-Hall, 1981.

Asimov on Astronomy, by Isaac Asimov. New York: Anchor Press, 1975. (In a life's work of more than 250 books, Isaac Asimov has produced dozens about space, astronomy, and scientific history. Rather than attempting to give a complete list of them here, we suggest that a visit to any bookstore will produce several informative volumes.)

Update on Space, edited by B. J. Bluth. Granada Hills, CA: National Behavior Systems, 1981.

The High Road, by Ben Bova. Boston: Houghton Mifflin, 1981.

Chariots for Apollo: A History of Manned Lunar Spacecraft, by Courtney Brooks, James Grimwood, and Loyd Swenson, Jr. Washington, DC: U.S. Government Printing Office, 1979.

The World in Space, edited by Ralph Chipman. Englewood Cliffs, NJ: Prentice-Hall, 1982.

The Challenge of the Spaceship, by Arthur C. Clarke. New York: Pocket Books, 1980. (Like Asimov, Clarke has written extensively on space. All his books are recommended as food for thought.)

Report on Planet Three, by Arthur C. Clarke. New York: Harper & Row (hardcover) and Signet Books (paperback), 1981.

Voices from the Sky, by Arthur C. Clarke. New York: Pocket Books, 1981.

Carrying the Fire: An Astronaut's Journeys, by Michael Collins. New York: Farrar, Straus & Giroux, 1974.

Space: The Final Frontier, by Heather Couper. New York: Crescent Books, 1980.

The View from Planet Earth, by Vincent Cronin. New York: Quill, 1983.

Space: The High Frontier in Perspective, by Daniel Deudney. Washington, DC: Worldwatch Institute, 1982.

Between Sputnik and the Shuttle: New Perspectives on American Astronautics, edited by Frederick C. Durant III. American Astronautical Society History Series, Volume 3, 1981.

Two Hundred Years of Flight in America, edited by Eugene M. Emme. American Astronautical Society History Series, Volume 1, 1977.

Inventing the Future, by Dennis Gabor. Baltimore: Pelican Books, 1963.

The Illustrated Encyclopedia of Space Technology, by Kenneth Gatland. New York: Harmony Books, 1981.

Migration to the Stars, by Edward S. Gilfillan. Washington, DC: Robert B. Luce Company, 1975.

Space Trek: The Endless Migration, by Jerome C. Glenn and George S. Robinson. Harrisburg, PA: Stackpole Books (hardcover); New York: Warner Books (paperback), 1983.

Enterprise, by Jerry Grey. New York: William Morrow & Co., 1979.

Apollo: Ten Years Since Tranquillity Base, edited by R. P. Hallion and T. D. Crouch. Washington, DC: Smithsonian Institution Press, 1979.

The Endless Space Frontier: A History of the House Committee on Science and Astronautics, by Kenneth Hechler. American Astronautical Society History Series, Volume 4, 1982. (This is an abridged version of the work: *Towards the Endless Frontier: History of the House Committee on Science and Technology, 1959–79,* published by the U.S. Government Printing Office, 1980.)

Colonies in Space, by Tom Heppenheimer. Harrisburg, PA: Stackpole Books (hardcover); New York: Fawcett (paperback), 1978.

Toward Distant Suns, by Tom Heppenheimer. Harrisburg, PA: Stackpole Books (hardcover); New York: Fawcett (paperback), 1980.

The Cosmic Chase, by Richard Hutton. New York: New American Library, 1981.

Outer Space: A New Dimension of the Arms Race, by Bhupendra Jasani. Cambridge, MA: Oelgeschlager, Gunn, and Hain, 1982.

The Space Shuttle Operator's Manual, by Kerry Joels and Gregory Kennedy. New York: Ballantine Books, 1982.

Handbook of Soviet Manned Space Flight and *Handbook of Soviet Lunar and Planetary Exploration,* by Nicholas L. Johnson. American Astronautical Society Science and Technology Series, Volumes 47 and 48, 1979 and 1980.

The New High Ground, by Thomas Karas. New York: Simon & Schuster, 1983.

Life in Space, by Dinah L. Moché. New York: A&W Visual Library, 1979.

Space Shuttle Guidebook, NASA publication, reproduced with permission and available through the National Space Institute, 1981.

Red Star in Orbit, by James Oberg. New York: Random House, 1981.

The Fertile Stars, by Brian O'Leary. New York: Everest House, 1981.

The High Frontier, by Gerard K. O'Neill; New York: William Morrow & Co., 1978 (hardcover); Bantam (paperback), 1983.

The Rocket Team, by Frederick I. Ordway III and Mitchell R. Sharpe. New York: Thomas Y. Crowell, 1979.

Battle for Space, by Kurtis Peebles. New York: Beaufort Books, 1983.

Global Talk, by Joseph Pelton. Rockville, MD: Sitjhoff & Noordhoff, 1981.

A Step Farther Out, by J. E. Pournelle. New York: ACE Books, 1979.

Living in Outer Space, by George S. Robinson. Washington, DC: Public Affairs Press, 1975.

The Cosmic Connection: An Extraterrestrial Perspective, by Carl Sagan. New York: Doubleday, 1973; Dell, 1975.

Cosmos, by Carl Sagan. New York: Random House, 1980.

Space Transportation Systems, 1980–2000, edited by Robert Salkeld, D. W. Patterson, and Jerry Grey. American Institute of Aeronautics and Astronautics, 1978.

Soviet Space Programs 1966–70; Soviet Space Programs 1971–75; Soviet Space Programs 1976–1980, by Charles S. Sheldon III, et al. Washington, DC: U.S. Government Printing Office, 1971, 1976, 1982. (This is a series of detailed reviews of the Soviet space program. The 1976–80 period will be covered in three volumes, only one of which has appeared at this date.)

The Third Industrial Revolution, by G. Harry Stine. New York: ACE Books, 1979.

Confrontation in Space, by G. Harry Stine. Englewood Cliffs, NJ: Prentice-Hall, 1981.

Gifts from Space, by L. B. Taylor, Jr. New York: John Day Company, 1977.

History of Rocketry and Space Travel, by Wernher von Braun and Frederick I. Ordway III. 3d ed. New York: Thomas Y. Crowell, 1975.

The Rockets' Red Glare: An Illustrated History of Rocketry Through the Ages, by Wernher von Braun and Frederick I. Ordway III. New York: Doubleday, 1976.

The Right Stuff, by Tom Wolfe. New York: Farrar, Straus & Giroux (hardcover); Bantam (paperback), 1980.

3

Training for a Space Career

"Knowledge is one. Its division into subjects is a concession to human weakness."
—Halford J. Mackinder

A Short Sermon

If you are serious about a career in space, and will not settle for playing around on the undemanding fringes of the field, then here is some important news. You must look forward to a lot of hard work. (We seem to be repeating this comment a lot, but that's only because it's so important).

Be patient with us for a moment while we preach at you. In this chapter, we will discuss training, mainly in terms of university programs and courses. But it will do no good to attend a university—any university—and assume that when you leave with a degree you have become "educated" and know all that you need to know. If you want a useful career in space, you are embarking on a lifetime of continuous learning. This does not mean it will not be enjoyable. Work on what interests you is usually intensely satisfying. But at the same time, there may be no jobs at all in the area that interests you most. The situation for space careers can be written as a simple formula:

Work opportunities = aptitude × training × enthusiasm × job openings.

If any one of the four variables on the right has a value of zero, it's not hard to see that the work opportunities, and hence your chance of finding a job, are also zero.

Aptitude is something you were born with. You can't change that. Job openings are beyond your control, reflecting the overall state of the economy and of the space program (though as we will point out in Chapter 5, you can increase the odds in your favor substantially by the correct approach to job hunting). Enthusiasm we assume you have already, or you would not be reading this book. The one area where you can make a huge difference in improving your chances is *training*.

Multiple Choice

"Education is hanging around until you've caught on."
—Robert Frost

"Education has really only one basic factor—a sine qua non—one must want it."
—G. E. Woodberry

We don't disagree with Mr. Frost and Mr. Woodberry, but in America today the word "education" unfortunately has two different meanings. There is the meaning when we refer to the need for lifelong education; and then there is the meaning implied when we say: "Where did she go for her education?" It is this second use that calls for a good deal more than "hanging around until you catch on," or wanting to learn, and it is the way you will be measured when you apply for a job. Like it or not, education to most employers is regarded as synonymous with schooling.

Why are employers so hung up on numbers of credits and grade-point averages? The problem is, no one can tell if you are *educated* in any real sense by looking at a piece of paper from a college. What they can tell is how many college credits you have, whether you graduated from high school, and so on. Is that education? No, it surely isn't— but it's all most employers have to go on apart from interviews, and it is the standard way you must expect to be evaluated in the job market. Almost everyone has a sneaking suspicion that credits and grades tell only part of the story, which is one reason for holding interviews. But to get as far as a job interview, you usually need plausible paper credentials. And those credentials, in large measure, tell a prospective employer little more than where you went to school, what you've experienced and studied. So the choices you make as to school and subject will prove to be highly important ones.

Let us examine some of the choices, and concentrate attention for the moment only on undergraduate education (graduate education has its turn in the section on Graduate Studies). First, take our word for it that you will find job openings in space-related fields a thousand times as available if you have a college degree than if you do not. This is true regardless of the school you attended, the subject that you studied, or the grades that you obtained.

Once you are inside an organization, quite different qualities decide how fast you will advance there. Joseph Fox, in his book *Executive Qualities* (Reading, Mass.: Addison-Wesley, 1976), argues that academic training is given far too much weight at the beginning: "What degree you have and from what university you received it doesn't make that much difference in the long run . . . A college degree will usually give the person beginning a career an edge over a co-worker who has no degree . . . But in the middle lev-

els, after the employee has proven excellence by performance, the degree issue is usually dropped."

We agree with this statement completely—but without a
degree you will unfortunately find it far harder to take the
first step of acceptance into the organization! If you can't
get in, you can't be promoted.

So you ought to do your utmost to gain a degree. But
which college should you attend? You have an enormous
field within which to make your choice. There are about
three thousand accredited colleges in the United States.
They range in size from schools with fewer than a hundred
students to campuses with more than forty thousand. They
are to be found in every state. Some schools are very expensive, others relatively inexpensive; some are highly se-
lective in entrance requirements, others are wide open.
Add in a variability in religious affiliations, plus the wide
range in courses offered and degree-program requirements,
and any hope of a simple summary of what is available
quickly disappears. The options are bewildering in their variety and number.

Not surprisingly, a whole industry has grown up around
the problem of college selection, and several hefty volumes
offer the most significant facts about each college, in directory form. If you do not already know these handbooks, we
recommend that you study one or two of them in your public library, or buy one. If you decide to buy, remember that
most volumes are updated annually, and that cost figures in
particular quickly become dated.

Suitable surveys include:

The College Handbook, published by the College Entrance Ex-
amination Board.

American Junior Colleges, published by the American Council on
Education (this covers two-year institutions only).

American Universities and Colleges, published by the American
Council on Education (this book is expensive, and you will
probably want to use a library copy rather than buying it).

Guide to American Graduate Schools, by Herbert B. Livesey and Harold Doughty, published by the Viking Press.

Peterson's Annual Guides to Graduate Study, Peterson's Guides, Princeton, New Jersey.

If you want to see the subject from a different viewpoint, try *The Underground Guide to the College of Your Choice,* by Susan Berman (published by New American Library). The book is a bit dated but explores such monumental questions as the male/female ratio of the student population, and the existence or lack of it of an underground student newspaper.

These books do nothing to narrow your choice; by showing how much is available, they probably enlarge it. However, we recommend that our comments be read hand in hand with one of these general guides. We lack the space or inclination to duplicate what can be found there, but that information (is tuition ten thousand dollars a year, or fifteen hundred dollars?) will be vital to a prospective student.

Picking a College

On the subject of choice of college, major, and courses, NASA takes the safe approach. In their guidance tips to aerospace careers, they pose and answer several questions as follows:

"What Colleges Does NASA Recommend?"

"NONE. NASA is not an accrediting agency, nor involved with evaluative activities qualifying it to make recommendations regarding this or that institution."

"What Courses Does NASA Recommend?"

"NONE. An undergraduate degree, from an accredited university, in science, mathematics, and/or engineering meets the basic entry requirement for most Aerospace Technology Specialties."

"What Jobs Will Be Available 5–10 Years in the Future?"

"No one knows for certain . . ."

These answers may be disappointing, but they are not unreasonable. As a government agency, NASA has to be very cautious about praising or damning any university or program; and as an agency that depends for its next year's budget entirely on actions by the President and by Congress, NASA cannot promise any particular job opportunities five or ten years from now. All the same, those answers don't help much when you are trying to decide on a college; and there is an additional suggestion in NASA's answers that if you don't take math and science you are out of your mind in seeking a space career.

We are not a government agency, so we are free to stick our necks out a lot farther when it comes to college choice. And unlike NASA, we were also free to conduct our own survey of the universities. (Any government agency wishing to seek survey responses from a sizable group of organizations or individuals must first obtain approval from the Office of Management and Budget, and that procedure is long and tedious. But we could write to anyone that we chose, and we did. Here is a rare example where the private individual with small resources has a big practical advantage over a multibillion-dollar government agency.)

The survey we performed was certainly not complete. We confined our inquiry to four-year colleges in the fifty states, the District of Columbia, and Puerto Rico. Generally speaking, we did not consider institutions with fewer than two thousand students, although we felt free to break this rule when our own experience suggested that a smaller college deserved consideration. We omitted some larger colleges that we judged crackpot in various ways (no, we won't say which ones; write to us if you want details—we prefer to avoid libel suits).

Most important, perhaps, we did not seek to concentrate only (or even mainly) on the stronger technical institutions. We know from much correspondence in preparing this book that many readers intend to obtain their academic training in the arts or humanities, and want to apply those somehow to a space career.

Our final list contained 514 universities, ranging from the

richest and most prestigious in the country to rather modest and little-known state schools. Arguing that different readers have different needs, financial resources, and preferences, we wrote to each college on the list.

In evaluating their replies, we gave high weight to several points, thus:

- Did we receive a knowledgeable response, suggesting that there are faculty and staff members who are already sympathetic to or active in space affairs?
- Are there, in addition to general courses of study, electives that involve some type of space study—social, philosophical, political, or technical?
- Was the idea of a text on space careers greeted with enthusiasm by the respondents? (Some colleges were more than enthusiastic—they said they couldn't wait to get their hands on the text, and they wished it were already available for students.)
- Does the college offer student financial assistance of some kind? Student needs and qualifications are so variable that we could not look for any particular program, but we believe it important that such programs exist in candidate colleges.

We did not look particularly for schools offering a "Bachelor of Space" degree, or anything else that directly correlated with space. We would probably discourage anyone from taking such an undergraduate degree. We believe that it is far better to obtain your training in some conventional subject, and add to that as much material relevant to space as you can find. This gives you much more freedom of choice in the job market.

Based upon their responses to our inquiries, we now offer a selection of colleges that we think make a good starting list. Obviously, not all these schools should be considered as equally good choices—for example, we have tried to recommend at least one college in every state. However, we believe that the opportunities for space-oriented college courses are rather better in California, Illinois, Massachusetts, New Jersey, New York, and Pennsylvania than in most other states. In evaluating a college for yourself, you should not let our suggestions too much reduce your choice; for example, tuition rates are normally far lower for in-state students, and that can make the difference between college and no college. But once you find a college you can afford, the most important thing is that you should feel intellectually comfortable in its environment. That depends on personal tastes as to campus size, layout, and location, as well as preferred companions and lifestyle. However, in addition to your personal questions, we suggest you ask the following questions concerning any college that you may be considering:

1) Are its entrance requirements close to your own qualifications? You don't want to be forced too far past your capacity, and you don't want to be bored. That's true for any career plan, not just a space-oriented one.

2) Does the school have a sizable science faculty? If it does not, think again. Even if you feel that your own strengths are not in science and you are seeking an arts degree, we urge you to take as many science courses as you can stand. When you embark on a space career, you are entering a technical field, and you want to understand as much of it as you can even if you do not specialize in that technology yourself. (This point is less relevant if you are proposing to start at a two-year college and then seek a transfer. In that case, the school to which you will transfer is the one where you should look for the science-faculty strength. We did not poll two-year colleges in our survey, not because we think they are a bad idea, but only because we had to keep the survey to a reasonable size.)

3) Can you *afford* to attend? If you do not have money yourself, what are the financial-assistance possibilities from the school, given your financial state and proposed field of study?

To answer this last question you almost certainly need to contact the school directly.

4) Does the college satisfy your built-in ideas as to how a college campus should appear? Not everyone covets ivy-covered walls, but most people have a mental image of a suitable college environment.

5) Is the college a member of the Universities Space Research Association? (See Chapter 4 for a description of this group.)

Suggestions for Colleges (alphabetical by state and within state)

Alabama

Jacksonville State University
Jacksonville, Alabama 36265

University of Alabama in
   Huntsville
P.O. Box 1247
Huntsville, Alabama 35807

University of South Alabama
Mobile, Alabama 36688

Alaska

University of Alaska
Fairbanks, Alaska 99701

Arizona

Arizona State University
Tempe, Arizona 85287

University of Arizona
Tucson, Arizona 85721

Arkansas

University of Arkansas at
   Fayetteville
Fayetteville, Arkansas 72701

California

California Institute of Technology
Pasadena, California 91109

California State University at
   Long Beach
6101 East Seventh Street
Long Beach, California 90840

California State University at
   Northridge
18111 Nordhoff Street
Northridge, California 91324

Golden Gate University
536 Mission Street
San Francisco, California 94105

Harvey Mudd College
Claremont, California 91711

San Diego State University
San Diego, California 92115

Stanford University
Stanford, California 94305

University of California at
   Berkeley
Berkeley, California 94720

University of California at Davis
Davis, California 95616

University of California at Irvine
Irvine, California 92664

University of California at Los
  Angeles
Los Angeles, California 90024

University of California at
  Riverside
Riverside, California 92502

University of California at San
  Diego
P.O. Box 109
La Jolla, California 92037

University of California at Santa
  Barbara
Santa Barbara, California 93106

University of the Pacific
Stockton, California 92211

University of Southern California
Los Angeles, California 90007

Colorado

Colorado School of Mines
Golden, Colorado 80401

Colorado State University
Fort Collins, Colorado 80523

United States Air Force Academy
USAF Academy, Colorado 80840

University of Colorado at
  Colorado Springs
Colorado Springs, Colorado 80907

University of Northern Colorado
Greeley, Colorado 80639

University of Southern Colorado
Pueblo, Colorado 81001

Connecticut

United States Coast Guard
  Academy
New London, Connecticut 06320

University of Connecticut
Storrs, Connecticut 06268

University of Hartford
200 Bloomfield Avenue
West Hartford, Connecticut 06117

Yale University
1502A Yale Station
New Haven, Connecticut 06520

Delaware

University of Delaware
Newark, Delaware 19711

District of Columbia

Georgetown University
Washington, D.C. 20051

George Washington University
Washington, D.C. 20006

Howard University
Washington, D.C. 20059

Florida

Embry-Riddle Aeronautical
  University
Regional Airport
Daytona Beach, Florida 32015

Florida Institute of Technology
Country Club Road
Melbourne, Florida 32901

Florida State University
Tallahassee, Florida 32306

Jacksonville University
Jacksonville, Florida 32211

University of Central Florida
Orlando, Florida 32816

University of Florida
Gainesville, Florida 32601

Georgia

Emory College
1380 Oxford Road
Atlanta, Georgia 30322

Georgia College
Milledgeville, Georgia 31061

Georgia (cont'd.)

Georgia Institute of Technology
Atlanta, Georgia 30332

Georgia State University
Atlanta, Georgia 30303

Southern Technical Institute
Marietta, Georgia 30060

University of Georgia
Athens, Georgia 30602

Hawaii

University of Hawaii at Manoa
Honolulu, Hawaii 96822

Idaho

University of Idaho
Moscow, Idaho 83843

Illinois

Bradley University
Peoria, Illinois 61625

Chicago State University
95th Street and King Drive
Chicago, Illinois 60628

DeVry Institute of Technology
3300 North Campbell Avenue
Chicago, Illinois 60618

Illinois Institute of Technology
Chicago, Illinois 60616

Northwestern University
Evanston, Illinois 60201

Southern Illinois University
Carbondale, Illinois 62901

University of Chicago
1116 East 59th Street
Chicago, Illinois 60637

University of Illinois
Urbana, Illinois 61801

Indiana

Butler University
Indianapolis, Indiana 46208

Indiana State University
Terre Haute, Indiana 47809

Indiana University at South Bend
South Bend, Indiana 46615

Indiana University-Purdue
    University at Indianapolis
Indianapolis, Indiana 46202

Purdue University
West Lafayette, Indiana 47907

University of Evansville
P.O. Box 329
Evansville, Indiana 47702

University of Notre Dame
Notre Dame, Indiana 46556

Iowa

University of Iowa
Iowa City, Iowa 52240

Kansas

Fort Hays Kansas State College
Hays, Kansas 67601

University of Kansas
Lawrence, Kansas 66045

Washburn University of Topeka
Topeka, Kansas 66621

Wichita State University
Wichita, Kansas 67208

Kentucky

Morehead State University
Morehead, Kentucky 40351

Murray State University
Murray, Kentucky 42071

University of Kentucky at
    Lexington
Lexington, Kentucky 40506

Louisiana

Louisiana Tech University
Tech Station
Ruston, Louisiana 71272

McNeese State University
Lake Charles, Louisiana 70601

Tulane University
New Orleans, Louisiana 70118

University of Southwestern
Louisiana
Lafayette, Louisiana 70501

Maine

University of Maine at Portland-
Gorham
96 Falmouth Street
Portland, Maine 04103

Maryland

Johns Hopkins University
Baltimore, Maryland 21218

United States Naval Academy
Annapolis, Maryland 21402

University of Baltimore
Baltimore, Maryland 21201

University of Maryland: College
Park Campus
College Park, Maryland 20742

Massachusetts

American International College
Springfield, Massachusetts 01109

Amherst College
Amherst, Massachusetts 01002

Boston College
Chestnut Hill, Massachusetts
02167

Boston University
Boston, Massachusetts 02215

Bridgewater State College
Bridgewater, Massachusetts 02324

Massachusetts Institute of
Technology
77 Massachusetts Avenue
Cambridge, Massachusetts 02139

Salem State College
Salem, Massachusetts 01970

Southeastern Massachusetts
University
North Dartmouth, Massachusetts
02747

Tufts College
Medford, Massachusetts 02155

University of Massachusetts
Amherst, Massachusetts 01003

University of Massachusetts at
Boston
Boston, Massachusetts 02125

Worcester Polytechnic Institute
Worcester, Massachusetts 01609

Michigan

Calvin College
Grand Rapids, Michigan 49506

Michigan State University
East Lansing, Michigan 48824

Michigan Technological University
Houghton, Michigan 49931

University of Michigan
Ann Arbor, Michigan 48104

University of Michigan-Dearborn
4901 Evergreen Road
Dearborn, Michigan 48128

Minnesota

Mankato State College
Mankato, Minnesota 56001

Moorhead State College
Moorhead, Minnesota 56560

St. Olaf College
Northfield, Minnesota 55057

Minnesota (cont'd.)

University of Minnesota
Minneapolis, Minnesota 55455

University of Minnesota
St. Paul, Minnesota 55108

University of Minnesota: Morris
Morris, Minnesota 56267

Mississippi

Mississippi State University
Mississippi State, Mississippi
39762

University of Mississippi
University, Mississippi 38677

Missouri

Central Missouri State University
Warrensburg, Missouri 64093

Northeast Missouri State
University
Kirksville, Missouri 63501

Northwest Missouri State
University
Maryville, Missouri 64468

Southeast Missouri State
University
Cape Girardeau, Missouri 63701

Washington University
St. Louis, Missouri 63130

Montana

Montana State University
Bozeman, Montana 59717

Nebraska

University of Nebraska
Lincoln, Nebraska 68508

University of Nebraska at Omaha
Omaha, Nebraska 68182

Nevada

University of Nevada: Las Vegas
Las Vegas, Nevada 89154

University of Nevada: Reno
Reno, Nevada 89557

New Hampshire

Dartmouth College
Hanover, New Hampshire 03755

Plymouth State College of the
University of New Hampshire
Plymouth, New Hampshire 03264

University of New Hampshire
Durham, New Hampshire 03824

New Jersey

Fairleigh Dickinson University:
Florham-Madison Campus
285 Madison Avenue
Madison, New Jersey 07940

Fairleigh Dickinson University:
Rutherford Campus
270 Montross Avenue
Rutherford, New Jersey 07070

Fairleigh Dickinson University:
Teaneck-Hackensack Campus
1000 River Road
Teaneck, New Jersey 07666

Monmouth College
West Long Branch, New Jersey
07764

New Jersey Institute of
Technology
323 High Street
Newark, New Jersey 07102

Princeton University
Princeton, New Jersey 08544

Ramapo College of New Jersey
505 Ramapo Valley Road
Mahwah, New Jersey 07430

Rutgers University—The State
University of New Jersey:
Livingston College
New Brunswick, New Jersey 08903

Stevens Institute of Technology
Hoboken, New Jersey 07030

New Mexico

Eastern New Mexico University
Portales, New Mexico 88130

University of New Mexico
Albuquerque, New Mexico 87131

New York

City University of New York
101 West 31st Street
New York, New York 10001

City University of New York:
York College
Jamaica, New York 11451

Colgate University
Hamilton, New York 13346

Cornell University
Ithaca, New York 14850

Fordham University
Bronx, New York 10458

Iona College
New Rochelle, New York 10801

Ithaca College
Ithaca, New York 14850

Niagara University
Niagara University, New York
14109

Polytechnic Institute of New York
333 Jay Street
Brooklyn, New York 11201

Rensselaer Polytechnic Institute
Troy, New York 12181

State University of New York at
Albany
1400 Washington Avenue
Albany, New York 12222

State University of New York at
Brockport
Brockport, New York 14420

State University of New York at
Buffalo
Buffalo, New York 14260

State University of New York
College at Oneonta
Oneonta, New York 13820

State University of New York
College at Potsdam
Potsdam, New York 13676

State University of New York,
Empire State College
Saratoga Springs, New York 12866

United States Military Academy
West Point, New York 10996

University of Rochester
Rochester, New York 14627

North Carolina

Duke University
Durham, North Carolina 27706

East Carolina University
Greenville, North Carolina 27834

University of North Carolina at
Charlotte
UNCC Station
Charlotte, North Carolina 28223

University of North Carolina at
Greensboro
Greensboro, North Carolina 27412

North Dakota

University of North Dakota
Grand Forks, North Dakota 58202

Ohio

Kent State University
Kent, Ohio 44242

Miami University
Oxford, Ohio 45056

Ohio Institute of Technology
1350 Alum Creek Drive
Columbus, Ohio 43209

University of Cincinnati
Cincinnati, Ohio 45221

Wittenberg University
Springfield, Ohio 45501

Wright State University
Dayton, Ohio 45435

Oklahoma

Cameron University
Lawton, Oklahoma 73501

Oklahoma State University
Stillwater, Oklahoma 74074

University of Oklahoma
Norman, Oklahoma 73069

Oregon

Oregon Institute of Technology
Klamath Falls, Oregon 97601

Southern Oregon State College
Ashland, Oregon 97520

University of Oregon
Eugene, Oregon 97403

Pennsylvania

Carnegie-Mellon University
5000 Forbes Avenue
Pittsburgh, Pennsylvania 15213

East Stroudsburg State College
East Stroudsburg, Pennsylvania
 18301

Lehigh University
Bethlehem, Pennsylvania 18105

Pennsylvania State University
University Park, Pennsylvania
 16802

Robert Morris College
Coraopolis, Pennsylvania 15108

Swarthmore College
Swarthmore, Pennsylvania 19081

University of Pennsylvania
Philadelphia, Pennsylvania 19104

Villanova University
Villanova, Pennsylvania 19085

Puerto Rico

Catholic University of Puerto Rico
Ponce, Puerto Rico 00732

University of Puerto Rico:
 Mayaguez Campus
Mayaguez, Puerto Rico 00708

Rhode Island

Brown University
Providence, Rhode Island 02912

Rhode Island College
Providence, Rhode Island 02908

University of Rhode Island
Kingston, Rhode Island 02881

South Carolina

The Citadel
Charleston, South Carolina 29409

Clemson University
Clemson, South Carolina 29631

University of South Carolina
Columbia, South Carolina 29208

South Dakota

University of South Dakota
Vermillion, South Dakota 57069

Tennessee

Middle Tennessee State University
Murfreesboro, Tennessee 37132

University of Tennessee at
  Knoxville
Knoxville, Tennessee 37996

Vanderbilt University
Nashville, Tennessee 37212

Texas

North Texas State University
Denton, Texas 76203

Rice University
Houston, Texas 77001

Southwest Texas State University
San Marcos, Texas 78666

Texas Tech University
Lubbock, Texas 79409

University of Texas at Arlington
Arlington, Texas 76019

University of Texas at Austin
Austin, Texas 78712

West Texas State University
Canyon, Texas 79016

Utah

University of Utah
Salt Lake City, Utah 84112

Utah State University
Logan, Utah 84322

Vermont

University of Vermont
Burlington, Vermont 05401

Virginia

University of Virginia
Charlottesville, Virginia 22903

Virginia Polytechnic Institute and
  State University
Blacksburg, Virginia 24061

Washington

Central Washington University
Ellensburg, Washington 98926

University of Puget Sound
Tacoma, Washington 98416

Washington State University
Pullman, Washington 99163

Western Washington State College
Bellingham, Washington 98225

West Virginia

Fairmont State College
Fairmont, West Virginia 26554

West Virginia University
Morgantown, West Virginia 26505

Wisconsin

University of Wisconsin—Eau
  Claire
Eau Claire, Wisconsin 54701

University of Wisconsin—
  Madison
Madison, Wisconsin 53706

University of Wisconsin—
  Milwaukee
2500 East Kenwood Boulevard
Milwaukee, Wisconsin 53201

University of Wisconsin—Stout
Menomonie, Wisconsin 54751

Wyoming

University of Wyoming
Laramie, Wyoming 82070

At this point, we think we hear cries of anguish. The reader is already in school—but at a college not on our list!

We don't regard this fact—alone—as cause for alarm. If your school is not on the list, it is in good company. We have also left out, for example, Harvard and Columbia, because although they are top universities, we preferred others in the same general area. We have also not included every college that belongs to the Universities Space Research Association (see Chapter 4), even though we believe that a good space-oriented education can be obtained at all those schools. Our selection is a highly personal one, and everyone's selection is likely to be different.

If you are attending a school that is *not* on the list, however, it is appropriate to begin asking questions:

Are you attending this particular university for personal reasons unrelated to your main career plans? (This happens quite often, when parents and friends are alumni, or when scholarships are restricted to the children of employees of certain companies, or from certain geographical locations.)

Are some of the faculty at your college knowledgeable of or interested in space research?

Do the faculty think of such research as a waste of money, or a waste of time?

Can the career-guidance department tell you who is hiring for jobs connected with space?

Are on-campus recruiting sessions held once or twice a year, at which aerospace companies or other high-technology companies are present?

Is there an amateur space-support group at the college?

If so, does the school make its facilities freely available for such a group's activities?

If your answers to these questions make you uneasy, then it is perhaps a good idea to look at other college choices.

Graduate Studies

It is harder to be helpful in this area, so we will be brief.

First, if you have already made it through an undergraduate program, your own college knows your talents and track record. They should be able to give you a much better idea of what to look for in graduate school than the general comments that we can offer. Second, it is not useful to evaluate graduate programs in the same overall way that one can comment on undergraduate programs. Student needs are far more specialized, and opportunities depend very much on the availability of first-rate individual faculty members. (It's well known that a large proportion of Nobel Prize winners did their graduate work under professors who were also Nobel Prize winners.) The first rule of graduate studies is to pick a really good adviser.

At the same time, the choice of school is terribly important. It is probably more critical to pick the correct graduate school than the correct undergraduate school. Although we said nothing about the relative prestige of different schools in discussing undergraduate programs, we have a different view at the graduate level. We urge you to try for the most prestigious graduate program that you can find and afford.

It is unfortunately also true that the most prestigious schools are the most expensive, so most people must find a balance between status and cost. At the same time, there are many chances to earn part or even all of the money you need by working while you study, within the college. The most painless funding, if you can get it, is through a fellowship. This normally does not require teaching or grading duties, though it may mean working on directed research in an area not central to your main graduate-study interest. Apart from the money earned, tuition costs will normally be waived as part of the fellowship benefits.

Teaching assistantships also provide supplemental income and waived tuition, but they will take a good deal of your time—often to the point where the student feels as though there is no time left for his or her own research work. And some professors undoubtedly regard graduate students as a convenient form of cheap labor to help with their own research interests. (It is an old student complaint that slavery has been abolished in this country, except in the graduate-study programs of our universities.) Again, we stress the importance of finding the right research adviser. Before you make a decision, be sure to talk to some of the other students, not just to the university counselors. The opinions of professors and the opinions of students are often wildly different.

A third method of funding graduate studies makes use of industrial study programs. A number of companies will allow employees a leave of absence to pursue a master's or a doctoral degree, and will pay tuition and partial salary while you do it. Naturally, there are usually a few strings attached. You will be required to maintain a satisfactory level of performance, and you will be expected to complete the study program in a specified time. Finally, there will often be an understanding that you will return for some period of time to work for the company that paid your way. In practice, most companies realize that keeping an employee who wants to leave is not in the company's own best interests, and they will try to hold you by offering positive inducements to stay. However, be sure you know what you are getting into before you take a program of this type— you may find that you are agreeing to pay everything back to the company unless you fulfill certain conditions for acceptable employment when the study period is over.

As a final comment, let us offer advice that most educators will regard as heretical. We do not think it is necessarily a good idea to pursue a graduate degree immediately upon completion of a bachelor's degree, and perhaps you should not pursue one at all.

This is particularly true if you want a somewhat unconventional career involving space. There is then a good chance that you will find no graduate degree program suitable to the field that you wish to pursue. For example, if you wish to be at the forefront of spaceborne computer development, you will be closer to the real action with Texas Instruments, Hewlett-Packard, or many other companies than at any university in the country. If you want to work on solar sails, you can do so just as well outside the college system as inside it. If your tastes run to the arts or entertainment implied by the space program, then almost certainly you should *not* be pursuing those interests at a university.

As we said earlier, education means more than a college degree. It may be that at the graduate-study level you will become educated faster in your chosen field outside the university system.

4

Getting
Involved

It is not necessary to have a space-related job in order to become involved in space-program activities and issues. Your participation can begin today, and it can occupy as much free time as you have available.

There exist in the United States several hundred organizations, all enthusiastic about space, and all dedicated to developing the country's and the world's space activities. These groups are of all sizes and special interests, from engineering to space colonization to planetary exploration. Some are political and encourage lobbying efforts on behalf of particular space-program missions; some are mainly scholarly, producing symposia, books, and journals about space; others provide general information about space-program activities, or arrange for members to be present at special events, such as space-shuttle launches or landings. Most rely heavily on assistance from amateur enthusiasts (remember that "amateur" has two meanings: a person who does something else for a living, and also a lover of the subject). And most have some form of newsletter, to tell their members what meetings will be held in the next few months, as well as to provide the bigger picture of what's going on in space.

Membership fees, and membership required qualifications, vary widely. But many membership fees and most additional contributions are tax-deductible (check with the Business Office of any society that interests you). Many groups require little or nothing in the way of technical expertise or prior space experience, nor do they require that you have any formal connection with the space program in order to join. The societies that do have entrance requirements are, predictably, the ones that place the highest emphasis on technical matters.

Taken together, these groups constitute what is sometimes termed the "space underground." As the phrase suggests, although independent, the societies will often work together to achieve some common goal. The combined membership is several hundred thousand, spread around this country and the world. In this chapter, we will describe the largest of them, ones that operate on a national or an international basis. Most of them possess local chapters, serving particular geographical areas. Some have special-interest groups within them, which may arrange their own meetings and newsletters. There are also many smaller groups, almost too numerous to list, and too changing in their interest and composition for an accurate description to be possible.

Finally, there are groups that emphasize some selected aspect of space development, such as space colonization, spaceborne astronomy, or women's space activities. We list many of these with addresses and telephone numbers at the end of this chapter.

We strongly recommend that the reader select from these lists the society or societies that most appeal to his or her individual taste, then contact them directly. There is no simpler way of gaining general information about what's going on in space, and no better way of becoming involved.

Space Organizations and Societies

THE AEROSPACE INDUSTRIES ASSOCIATION (AIA)

Like several other organizations in the list that follows, the Aerospace Industries Association has as its members not individuals, but other organizations; in this case, the AIA is a trade organization for companies in the United States that are engaged in research, development, and manufacture of aerospace systems. This includes manned and unmanned aircraft, missiles, space launch vehicles and spacecraft, propulsion, guidance and control units, and much of the airborne and ground-based equipment needed in the operation of flight vehicles.

Although AIA is not a direct source of jobs, or of information concerning jobs for individuals, it provides a valuable source of data to tell what is going on in the aerospace field. For example, it runs surveys of aerospace employment, by state. Those surveys are available to the general public, not merely to the AIA company members.

The association has a long history, going back to the Manufacturers Aircraft Association formed in 1917. Its present name was adopted in 1959, and today its forty-eight member companies include all the significant aerospace corporations (see the membership list on page 79). AIA also has eight affiliate organizations, which are not U.S. aerospace companies but which share many of their interests.

AIA is also itself a member of an international council for aerospace activities, ICCAIA (the International Coordinating Council of Aerospace Industries Associations). Through this affiliation, AIA has good contacts with the aerospace industries of many countries, including Belgium,

Canada, Denmark, France, Germany, Italy, Japan, Spain, Sweden, the Netherlands, and the United Kingdom. Note that although the AIA will *not* serve as a specific reference source for individual inquirers, it generates and makes available much general information that assists in employment searches—for example, AIA's Aerospace Research Center publishes the annual "Aerospace Facts and Figures"; a semi-annual employment survey; and other statistical reviews of the state of aerospace activities. In addition, the AIA annual report is itself a very good overview of the condition of the U.S. aerospace industry. It covers contract awards and contract backlog summaries, a description of AIA projects and publications for the year, and summaries of other matters such as international technology transfer, NATO activities, and international cooperation in aerospace affairs.

AIA member organizations include the following companies:

Abex Corporation

Aerojet-General Corp.

Aeronca, Inc.

Avco Corporation

Aluminum Corporation of America

The Bendix Corp.

The Boeing Company

CCI Corporation
—the Marquardt Company

Colt Industries, Inc.
—Chandler Evans, Inc.
—Menasco, Inc.

Criton Corporation

E-Systems, Inc.

FMC Corporation

Garrett Corporation

Gates Learjet Corp.

General Dynamics Corp.

General Electric Company

BF Goodrich Company

Goodyear Aerospace Corp.

Gould, Inc.

Grumman Corporation

Hercules Incorporated

Honeywell, Inc.

Howmet Turbine Corp.

Hughes Aircraft Company

IBM Corporation
—Federal Systems Division

ITT Telecommunications & Electronic Company—North America

ITT Aerospace

ITT Avionics Division
ITT Defense Communications
   Division
ITT Gilfillan
Lear Siegler, Inc.
Lockheed Corporation
Martin Marietta Aerospace
McDonnell Douglas Corp.
Northrop Corporation
Parker Hannifin Corp.
Pneumo Corporation
   —Cleveland Pneumatic Co.
   —National Water Lift Co.
Raytheon Company
RCA Corporation
Rockwell International Corp.
Rohr Industries, Inc.

The Singer Company
Sperry Corporation
Sundstrand Corporation
Teledyne CAE
Textron, Inc.
   —Bell Aerospace
   —Bell Helicopter
   —Dalmo Victor Operations
   —HR Textron, Inc.
Thiokol Corporation
TRW, Inc.
United Technologies Corp.
Vought Corporation
Western Gear Corp.
Westinghouse Electric Corp.
Wyman-Gordon Company

Affiliates include: Air Carrier Service Corporation, Associated Aerospace Activities, Inc., Aviquipo, Inc., British Aerospace, Inc., Commerce Overseas Corporation, Eastern Aircraft Corporation, National Credit Office, Inc., and U.S. Aviation Underwriters, Inc.

Additional information concerning the organization can be obtained from AIA headquarters, at 1725 De Sales Street N.W., Washington, D.C. 20036 (Telephone: 202-429-4656).

AEROSPACE MEDICAL ASSOCIATION

This is another active group with somewhat specialized aviation and space interests. In this case, the association concentrates on problems of space medicine, life sciences, and environmental sciences. It produces a monthly journal, *Aviation, Space, and Environmental Medicine,* and provides contact with physicians, life scientists, bioengineers, and medical specialists working in basic medical research and in its clinical applications. It holds an annual scientific meeting on aerospace medical matters, provides special recognition and awards for noteworthy contributions to aviation, space, or environmental medicine, and operates a continuing professional education program. It is a long-standing member of the International Astronautical Federation.

Membership is open to:

(a) college graduates or commissioned armed services members working in the fields of space medicine, aeronautics, astronautics, undersea medicine, or environmental health
(b) scientists, engineers, and technicians concerned in those same fields
(c) teachers and research workers in these fields
(d) licensed physicians
(e) flight medical officers and nurses, aviation medical examiners, submarine or diving medical officers, flight surgeons, and aviation medical directors

Annual membership dues are sixty-five dollars for a full member, and forty dollars for a student or intern member. Membership includes a subscription to the journal. Additional information concerning the organization may be obtained from:

> Aerospace Medical Association
> Washington National Airport
> Washington, D.C. 20001
> (Telephone: 703-892-2240)

THE AMERICAN ASTRONAUTICAL SOCIETY (AAS)

This is one of the oldest (established 1954) and best organized of the space organizations. It is also one of the most technically sophisticated, with a series of books on various space issues that is unmatched in the field. The subjects covered range from highly technical volumes on astrodynamics or guidance and control to general histories of manned flight and projections of long-term space futures.

The society's stated goals run thus: "The AAS is dedicated to the advancement of the astronautical sciences and spaceflight engineering and encouragement of the astronautic arts. The goals of the Society are furthered by the exchange of ideas and information throughout the world space community, through the organization of national and local meetings and symposia and publication of our archival journal, newsletter, and books."

The "archival journal" referred to above is a specialist volume, and difficult reading to nonmathematicians and nonengineers; however, the newsletter contains general information, and the published books are over a hundred volumes that span subjects from space safety to space shuttle, and from guidance and control to the first two hundred years of manned flight.

National meetings are held at least twice a year, one of these always being the Goddard Memorial Symposium in Washington, D.C. In addition to technical coverage of space matters, these meetings feature well-organized and well-attended history sessions, providing fascinating inside information on the early development of the space age.

Membership information and benefits are described as follows:

"The Society is composed of five grades of individual membership: STUDENT MEMBERS shall be interested in astronautics and be registered students at recognized educational institutions.

"AFFILIATE MEMBERS shall be at least 18 years of age and have demonstrated an interest in astronautics.

"MEMBERS shall have a personal involvement in the field of astronautics and have four years of training and/or work experience related to astronautics.

"SENIOR MEMBERS shall have engaged in professional work in astronautics or related fields for not less than ten years and have made significant contributions.

"FELLOWS shall be persons of national reputation in astronautics who have demonstrated an interest in the Society. Candidates are nominated and elected annually by the active Fellows with the approval of the Board of Directors.

"Current membership fees are as follows: Student Members, $15; Members and Affiliate Members, $35; Senior Members and Fellows, $60. Fees for new members are discounted at National Meetings and special events.

"Benefits of membership include:

- Formal and informal association with peers in astronautics;
- Participation in local, national and international meetings at reduced registration fees;
- Subscription to the Journal of the Astronautical Sciences and the AAS Newsletter;
- A 25% discount on all AAS proceedings and technical publications;
- An opportunity for individuals and corporations to present new and advanced ideas, concepts and programs in major forums and to support ongoing programs in space."

Of less importance to many members, but of great possible relevance if you are looking for a job in the space field, is the fact that the AAS also admits corporate members. These are companies who are actively involved in the space business. Their representatives attend all national meetings, and are usually highly accessible. This type of interaction is an unrivaled method for finding out about upcoming projects, and for getting an inside feel for what work at a particular company is really like.

For more information, contact the AAS Business Office, 6060 Duke Street, Alexandria, Virginia 22304 (Telephone: 703-751-7721).

THE AMERICAN ASTRONOMICAL SOCIETY (AAS)

This society, like the American Astronautical Society, uses the initials AAS—a fact that causes a good deal of confusion and transfers of mail between the societies.

The objectives of the two groups are quite different. The American Astronomical Society is the major organization of professional astronomers in the United States, Canada, and Mexico. Its interest is space—but in the traditional sense of astronomy, rather than in the recent senses of direct exploration, travel, and exploitation. Thus, its members are mainly physicists, mathematicians, geologists, astronomers, and engineers, most with strong technical backgrounds.

The society holds two general meetings each year, but additional joint meetings with other societies are common, to provide good cross-fertilization between astronomy and related disciplines. In view of the diversity of modern astronomy, specialized subdivisions have been organized in the areas of planetary sciences, solar physics, dynamical astronomy, high-energy astrophysics, and historical astronomy.

The AAS publishes four main journals for dissemination of research results. They are: *The Astrophysical Journal, The Astronomical Journal, The Bulletin of the American Astronomical Society,* and *The American Astronomical Society Photo-Bulletin.*

There is also a quarterly newsletter, describing society activities, giving information about federal agencies, and providing general news of interest to astronomers.

To extend general awareness of astronomy, the society conducts a lecture program, within which about a hundred visiting lectures a year are given at colleges that have little or no access to professional astronomy. In these, the Harlow Shapley Visiting Lectureships, AAS members provide information to students, faculty, and the general public. Additional public-education efforts include films and brochures. Internationally, the society provides the mechanism by which the United States participates in the International Astronomical Union.

The job services of the society include the publication of a monthly job register, available to all members upon request. A job center is also operated at each of the general meetings, and a Candidates Register of brief résumés is maintained for use by potential employers.

Membership Categories are as follows:

Junior membership is open to any person seriously interested in the advancement of astronomy and related fields, of twenty-eight years of age or younger. Full-time students pursuing a degree in astronomy or physics are entitled to junior membership upon furnishing proof of their full-time student status.

Associate membership is open to any person seriously interested in the advancement of astronomy or related fields.

Full membership is open to a person deemed capable of preparing an acceptable scientific paper upon some subject of astronomy or related branch of science. Full members normally have received a Ph.D. degree or equivalent.

Junior and associate members have all the benefits of full members, except that they may not serve on the nominating committee or as officers or councilors. Two full members must sign any nomination form for membership in the society, and must be familiar with the work of the party nominated.

Annual dues and subscriptions for 1983 (all tax-deductible) are as follows:

Junior members: eighteen dollars; associate members: fifty-four dollars; full members: fifty-four dollars; junior/ Society of Physics student member: twenty-four dollars; supporting contribution (optional): twenty-five to fifty dollars.

All membership dues include a subscription to the *AAS Newsletter* and to *Physics Today*.

Additional information concerning the society may be obtained from the national headquarters, at this address:

American Astronomical Society
1816 Jefferson Place, N.W.
Washington, D.C. 20036
(Telephone: 202-659-0134)

THE AMERICAN INSTITUTE OF AERONAUTICS AND ASTRONAUTICS (AIAA)

This is the largest of the technical societies wholly devoted to the advancement of aeronautics and astronautics. It grew out of the old American Rocket Society, and now has these stated goals: "To advance the arts, sciences, and technology of aeronautics and astronautics, and to nurture and promote the professionalism of those engaged in these pursuits." Also, "to serve the needs and professional interests of members, [and] improve public understanding of the profession and its contributions . . ."

The society supports six technical specialty groups and more than fifty technical committees. It is geographically divided into six national regions, which in turn contain sixty-six local sections. There are also 120 student branches in colleges and universities across the country, and student activities play a large part in society affairs and priorities. One of the standing committees is specifically devoted to Career Environment, another to Student Activities.

The AIAA is very active in national specialist conferences, and organizes about thirty of these each year in different parts of the United States. These meetings offer attendees an excellent opportunity to examine industrial and government exhibits, and to make field trips to aerospace plants and laboratories.

The national headquarters is in New York (the address is given later), but the society regards its sixty-six local sections as the "nerve centers" of the AIAA. Each has its own offices, its own program of activities, and its own lecture series and professional education opportunities. The AIAA's sixty-six sections are in the following locations (addresses for local contact points and identification of the section nearest to you are available from the national headquarters):

Region I (Northeast USA)

Allegheny/Pittsburgh
Baltimore
Blue Ridge
Central New York
Central Pennsylvania
Connecticut
Delaware

Greater New York
Greater Philadelphia
Hampton Roads
Long Island
National Capital
New England
Niagara Frontier

Northeastern New York
Northern New Jersey
Princeton
Southern New Jersey
Southern Tier

Region II (Southeast USA)

Alabama/Mississippi
Atlanta
Cape Canaveral
Carolina

Central Florida
Greater New Orleans
Northwest Florida

Palm Beach
Tampa Bay
Tennessee

Region III (East Central USA)

Central Indiana
Columbus
Dayton-Cincinnati

Illinois
Michigan
Northern Ohio

St. Joseph Valley
Wisconsin

Region IV (Southwest USA)

Albuquerque
Central Texas
Holloman-Alamo-
    gordo

Houston
Inland Missile Range
North Texas

Oklahoma
Southwest Texas

Region V (Great Plains and Rocky Mountain States)

Iowa
Rocky Mountain
St. Louis
Twin Cities

Wichita

Region VI (Western States)

Antelope Valley	Phoenix	Tucson
Arrowhead	Point Lobos	Utah
China Lake	Sacramento	Vandenberg
Los Angeles	San Diego	Ventura Pacific
Orange County	San Francisco	
Pacific Northwest	Santa Barbara	

Membership grades and requirements are as follows:

Associate members are persons who have an interest in the development or application of aeronautics or astronautics.

Members are persons who have acquired a professional standing in the practice of the arts, sciences, or technology of aeronautics or astronautics. Applicants for member grade shall have achieved a Bachelor degree in science or engineering, or equivalent qualifications through professional practice. Applicants shall furnish the names of two references, at least one being a senior member or higher.

Senior members are persons who have demonstrated a successful professional practice in the arts, sciences, or technology of aeronautics or astronautics for at least eight years (which may include engineering and science study programs). Senior members may be nominated by any member in good standing. Nominees must be members of the AIAA and shall be recommended by four persons, two of whom shall be senior members or higher.

Associate fellows shall be persons who have accomplished or been in charge of important engineering or scientific work or who have done original work of outstanding merit, or who have otherwise made outstanding contributions to the arts, sciences, or technology of aeronautics or astronautics. . . . There will be a maximum of one associate fellow for each 150 voting members of the AIAA. Associate fellows must be AIAA members, meet the requirements for the senior member grade, have twelve years professional experience, and be recommended by four persons, three of whom shall be associate fellows or higher.

Fellows and honorary fellows are the two highest grades of membership and are achieved only through nomination from the membership and subsequent election by the board of directors.

Publications *Aeronautics and Astronautics* is a monthly magazine that goes to every member. It discusses trends in the space program, both domestic and foreign, and reviews topical issues on the government, military, and commercial aerospace fronts. It is also the principal channel for communications with members, and the news section of this magazine provides information on national and local AIAA activities. *The AIAA Bulletin* is a section of *Aeronautics and Astronautics,* containing program information, meeting calendars, calls for papers, abstracts, and schedules for meetings.

The society also publishes a number of important archival journals in the aerospace field, including the *AIAA Journal,* the *Journal of Aircraft,* the *Journal of Energy,* the *Journal of Guidance, Control and Dynamics,* and the *Journal of Spacecraft and Rockets.* All these publications are of specialist rather than of general interest.

The AIAA also publishes a series of hardcover books on topics in astronautics and aeronautics, and assessments of topical aerospace matters, such as space transportation systems, solar-system exploration, supersonic-aircraft development, and global aspects of space technology. These are broad gauge in approach and generally accessible to the general reader.

Two other areas of particular AIAA activity should be mentioned. The society emphasizes finding jobs in the aerospace field, and has two handbooks designed for this purpose and available to members. These are *The AIAA Employment Workshop Handbook,* and *Job Hunting: The*

Seven Steps to Success. In areas of high unemployment, the society also sets up employment workshops through local sections. For student members, the AIAA also publishes a special student journal, with articles and advice on job selection and job hunting.

Second, the society is active internationally. It is a member of the International Astronautical Federation, and of the International Council for the Astronautical Sciences, and cosponsors numerous international meetings. It also works closely with the United Nations Committee on Peaceful Uses of Outer Space (COPUOS), has foreign members on the technical committees, and permits foreign companies to join the society as corporate members. Additional information on the AIAA and its activities may be obtained from the following sources:

> AIAA National Headquarters
> American Institute of Aeronautics
>   and Astronautics
> 1290 Avenue of the Americas
> New York, NY 10104
> (Telephone: 212-581-4300)
>
> AIAA Western Office
> 9841 Airport Boulevard
> Suite 800
> Los Angeles, CA 90045
> (Telephone: 213-670-6642)
>
> AIAA Washington Office
> 1625 Eye Street, N.W.
> Washington, DC 20006
> (Telephone: 202-785-0293)

THE AMERICAN SOCIETY FOR AEROSPACE EDUCATION (ASAE)

Founded in 1976, this organization is wholly dedicated to aviation and space education. It is nonprofit, independent, and not aligned with any industrial or governmental group. It is the U.S. representative on the seventy-two-nation International Council for Aerospace Education, and also sponsors the National Council for Aerospace Education and the aerospace activities of the National Councils of Elementary Teachers, Secondary Teachers, and University Educators.

Given this emphasis, it is no surprise that the ASAE's principal organ of publication, the bimonthly *Aviation/Space* magazine, has strong tutorial elements. It contains the fullest and most complete summaries of publications in the space and aviation field, including lists of free materials available from NASA and other government and nongovernment agencies. Tutorial articles on subjects such as the space shuttle, advanced air transport, aerospace careers, U.S. space policy, and planetary exploration appear regularly. Each issue contains a section on education, and another on resources—the latter being a review of all types of information, materials, and projects having to do with aviation and space.

Society membership is open to all, with annual dues (tax-deductible) of twenty-five dollars.

For additional information, contact ASAE headquarters at:

The American Society for Aerospace Education
1750 Pennsylvania Avenue, N.W., Suite 1303
Washington, DC 20006
(Telephone: 202-347-5187)

THE AMERICAN SOCIETY OF AEROSPACE PILOTS (ASAP)

This organization is designed mainly for qualified pilots. Full members must have a Commercial Pilot Certificate or an FAA ATP, plus considerable flight experience. These requirements may be waived at the discretion of ASAP's executive board. Full members must also participate in a minimum of twelve days of society-approved activities each year. Dues range from fifty dollars a year for a full member to fifteen dollars for a student associate member. The following description is taken from the society's general brochure:

The Organization   The American Society of Aerospace Pilots (ASAP) is a not-for-profit corporation dedicated to the promotion of routine commercial space flight and to the education of the public about the benefits of the commercial utilization of space. ASAP also devotes considerable effort to informing the public, and important decision-makers, about the contribution of the aerospace industry to our general welfare.

In addition, ASAP offers its members practical courses on all aspects of flying in space and, specifically, on space-shuttle operations. Much of this material is drawn from NASA astronaut-training courses.

Special VIP tours of various aerospace facilities (including trips to space-shuttle launches and landings), chapter meetings with noted guest speakers, and seminars on space operations are also benefits of ASAP membership.

Background   With commercial space flight becoming a virtual certainty in the future, a group of interested United Airlines pilots investigated the developments and possibilities in this exciting new field. After much research, it was found that there is widespread interest within the professional pilot group and the general public about the opportunities offered by commercial space operations. In fact, many expressed a desire to enroll in a Basic Spaceflight Ground School.

After considering various approaches, the American Society of Aerospace Pilots was formed as the vehicle to accelerate the evolution of commercial space operations and to prepare pilots to operate commercial spacecraft.

The American Society of Aerospace Pilots (ASAP) was incorporated in Illinois in September of 1981 as a not-for-profit, educational corporation.

Structure ASAP is a national organization with membership open to any interested person having a desire to support the society and its goals of educating its members and the public about the useful and beneficial aspects of the routine commercial use of space.

The structure of ASAP consists of a national headquarters in the Chicago area, and various chapters located in major cities in the United States.

Training Courses ASAP offers its members training courses on various aspects of space flight. The courses are designed for individual study augmented by seminars, lectures, and field trips.

The Basic Spaceflight Ground School Program is drawn from NASA astronaut-training material. The material is presented in a form familiar to most pilots and stresses operational space-flight operations.

For more information, contact ASAP headquarters, 1305 Remington Road, Suite I, Schaumburg, Illinois 60195 (Telephone: 312-884-7878).

THE BRITISH INTERPLANETARY SOCIETY (BIS)

This may look like the odd man out in a list of U.S. space groups. We include it here for three reasons. First, this society has been active for so long (it is celebrating its fiftieth anniversary) and has been an important influence on so many major figures in space development, worldwide, that its omission would be inexcusable.

Second, almost half of the BIS members live and work in this country, even though its headquarters and its officers are to be found in the United Kingdom.

Third, the bimonthly magazine *Spaceflight* and the monthly *Journal of the British Interplanetary Society* are justly famous, one as a general survey of what is going on in space, the other as a technical organ in which many important ideas have seen their first publication. For example, in 1938–39 the society produced a design for a multistage moon rocket that strongly resembles that used for the Apollo Program. In 1950, Arthur Clarke published in the BIS *Journal* the idea of an electromagnetic launch mechanism, that would much later be rediscovered and renamed a "mass driver." In 1952–53, the society produced the design for a reuseable launch vehicle, a "space shuttle." In 1978, Project Daedalus was completed by BIS members. This is the most complete analysis ever performed of an *interstellar* spacecraft, capable of traveling to the nearest stars in half a century. Most recently, the *Journal* has published articles on large orbital ring structures, matter-antimatter propulsion systems, and on planetary terraforming.

The objectives of the BIS are:

1. To promote the advancement of space research, technology, and applications.
2. To serve the general community by the interchange and dissemination of technical and other information by means of lectures, symposia, visits, and publications.
3. To promote the work of those professionally engaged in space research, space technology, and allied subject-areas.
4. To discuss national and international activities in space and to formulate forward-looking policies for the advancement of space exploration and utilization.

Membership dues, payable either in British pounds or U.S. dollars, are as follows:

Member under the age of 18	$35.00
Member, 18 to 20	$40.00
Members, 21 years and over	$45.00
Associate Fellows	$50.00
Fellows	$50.00

Membership includes a subscription to *Spaceflight*. Subscription to the *Journal* is an additional forty-four dollars per year.

Additional information about the society may be obtained by writing to:

The Executive Secretary
The British Interplanetary Society
27/29 South Lambeth Road
London SW8 1SZ
England

DELTA VEE, INC.

This group began life as a special activity of the American Astronautical Society named the Viking Fund. It was responsible for raising funds by private contributions, to keep receiving and processing data from the Viking spacecraft on the surface of Mars. The effort was very successful, and the funds raised were presented to NASA as an expression of public support for the U.S. space program.

Following the successful operation of the Viking Fund, Delta Vee was formed in 1980. Its purpose is to continue fund-raising programs for space, and to encourage space activism by private citizens and students. It encourages the idea of *creation* of a job in the space program by an individual's personal efforts.

Delta Vee projects include regular workshops and presentations on the subject of "Careers In Space: Your Guide to the Future." The organization has also established special fund-raising programs for private support for a Halley's Comet mission, and for the search for extraterrestrial intelligence. Of special interest to readers of this book, Delta Vee also runs a job-referral service, matching job-seekers to corporations with staff needs.

For more information, contact Delta Vee headquarters at 456 El Paseo de Saratoga, San Jose, California 95130 (Telephone: 408-370-0466).

THE HYPATIA CLUSTER

This group, founded in 1981, is the first organization established specifically to encourage women's involvement in space exploration. It is a nonprofit, educational organization, named after the renowned woman astronomer and mathematician Hypatia, who lived and taught at the beginning of the fifth century. The group's statement of purpose runs as follows: "The Hypatia Cluster functions to promote public awareness of the importance of space exploration, and particularly to recognize and encourage the participation of women in space science, in technology, and in space activities."

Despite the emphasis on women's involvement in space activities, the cluster is not for women only. This is made clear by its statement of specific objectives, to be achieved over the next five years and including roles for both men and women:

"The Hypatia Cluster will be recognized (by organizations, institutions, information networks, and individuals seeking information about space and women) and function as a popularly supported organization which:

1) promotes increased public awareness of the importance of space exploration and development;

2) promotes increased public participation in activities supporting space exploration and development;
3) encourages young women and men to enter space science, engineering, and related careers;
4) promotes public awareness of the specific contributions of today's "pioneers"—the women and men who are working in space sciences, aerospace professions, and space advocacy; and
5) places special emphasis on encouraging and recognizing the past, present, and future participation of women in space sciences, aerospace professions, and space advocacy."

The group is particularly involved in educational programs and offers advice, action programs, and educational sessions for students interested in preparing themselves for a space career. It is also the West Coast Coordinator for the National Ad Hoc Committee on Women and Space.

Dues (tax-deductible) for 1983 are as follows:

Student member	$5.00
Associate member	$10.00
Supporting member	$20.00
Founding member	$35.00
Lifetime member	$500.00

For additional information on the Hypatia Cluster, contact Amy Marsh or Marita Dorenbecher at the following address:

Hypatia Cluster
1724 Sacramento Street, Suite 200
San Francisco, CA 94109
(Telephone: 415-552-0141)

THE INSTITUTE OF ELECTRICAL AND ELECTRONICS ENGINEERS (IEEE)

This is a giant organization, with 230,000 members—the largest technical society in the world, celebrating its centennial in 1984. There is strong emphasis on educational activity, with 470 student branches and nearly 40,000 student members. All members receive two general society publications, *Spectrum* (an electrical and electronics magazine) and *The Institute* (a monthly newsletter). In addition, IEEE affiliates publish scores of specialized journals and transactions.

The general society objectives are scientific and educational, "directed toward the advancement of the theory and practice of electrical engineering, electronics, computer engineering and computer sciences and the allied branches of engineering and the related arts and sciences"; and professional, "directed toward the advancement of the standing of the members of the professions it serves." The society is international, with six of its ten regions in the United States, Region 7 in Canada, Region 8 in Europe, Africa and the Middle East, Region 9 in Latin America, and Region 10 covering Asia and the Pacific.

With so large a society, the best way to think of its activities is in terms of its thirty-one separate technical components. Each of these is itself a substantial society, with its own programs and publications. Each technical society also has its own membership fee and publication charges, in addition to the general IEEE membership dues.

The technical societies that make up the IEEE, together with their separate (1983) annual dues, are as follows:

Acoustics, Speech, and Signal Processing	$ 7.00
Aerospace and Electronic Systems	15.00
Antennas and Propagation	8.00
Broadcast, Cable, and Consumer Electronics	5.00
Circuits and Systems	12.00

Communications	15.00
Components, Hybrids, and Manufacturing Technology	9.00
Computer	10.00
Control Systems	12.00
Education	8.00
Electrical Insulation	12.00
Electromagnetic Compatibility	7.00
Electron Devices	5.00
Engineering Management	10.50
Engineering in Medicine and Biology	14.00
Geoscience and Remote Sensing	7.00
Industrial Electronics	10.00
Industry Applications	8.00
Information Theory	8.00
Instrumentation and Measurement	5.00
Magnetics	7.00
Microwave Theory and Techniques	8.00
Nuclear and Plasma Sciences	9.00
Power Engineering	10.00
Professional Communications	10.00
Quantum Electronics and Applications	15.00
Reliability	7.00
Social Implications of Technology	6.00
Sonics and Ultrasonics	7.00
Systems, Man, and Cybernetics	11.00
Vehicular Technology	10.00

In addition to the above dues, there are also general IEEE entrance fees and annual dues, variable by geographic region as follows:

	Entrance Fee	1983 Dues
United States	$15.00	$61.00
Canada	15.00	55.00
Europe, Africa, Middle East	15.00	55.00
Latin America	15.00	48.00
Asia and Pacific	15.00	48.00

Reduced dues are available for student members, or for members living on low incomes (written certification is needed for this).

From the point of view of space activities, it is the Aerospace & Electronic Systems Society that is most likely to be of interest to readers of this book. However, several others are worth exploring by the nonengineer, particularly those for Social Implications of Technology and for Education.

Since the society is large and its dues and organization structure are so complex, we recommend that interested readers obtain a membership brochure that goes into these subjects in detail. This is obtainable from:

> IEEE Service Center
> 445 Hoes Lane
> Piscataway, NJ 08854
> (Telephone: 201-981-0060)

For more information on the society in general, or on its aerospace-related functions, contact IEEE headquarters:

> IEEE, Inc.
> 345 East 47th Street
> New York, NY 10017
> (Telephone: 212-705-7900)

THE INTERNATIONAL ASTRONAUTICAL FEDERATION (IAF)

This organization is included mainly to avoid confusion. It is an active group, and holds many excellent international meetings. However, it is a society whose members are *societies*, not individuals. Founded in 1950, it contains sixty-four member organizations, drawn from thirty-six countries. Thus, you cannot become a member of the IAF—but you can join a society that is itself a member. This will assure that you have full information concerning IAF activities.

The IAF's Constitution states its goals: to foster the development of astronautics for peaceful purposes, to encourage the widespread dissemination of technical information, to stimulate public interest in space flight through the major media of mass communication, to encourage astronautics research, to convoke congresses and scientific meetings, and to cooperate with other organizations in all aspects of natural, engineering, and social sciences related to astronautics and the peaceful uses of outer space. The IAF holds an annual International Astronautical Congress, held each year in a different country.

In 1960, the IAF created the International Academy of Astronautics (IAA) and the International Institute of Space Law (IISL). These are autonomous organizations, though they cooperate closely with the IAF. Both the IAA and the IISL admit individual members. However, these members are all elected and neither group is therefore open to general membership. The academy publishes *Acta Astronautica,* a professional astronautics journal. The institute organizes an annual colloquium on the Law of Outer Space, held simultaneously with the IAF Congress. The IAA has about 550 members, the IISL about 350.

The North American organizations who are members of the IAF include the American Astronautical Society, the Aerospace Medical Association, the American Institute of

Aeronautics and Astronautics, the Canadian Aeronautics and Space Institute (Ottawa), and the Sociedad Mexicana de Estudios Interplanetarios (Mexico City). Each of these groups can tell you about upcoming IAF meetings. For general information about the IAF, however, it is better to write to their headquarters, at:

> The International Astronautical Federation
> 250 Rue Saint-Jacques
> 75005 Paris, France

THE L-5 SOCIETY

Background This society was founded in 1975, with the initial sole and ambitious purpose of placing a space colony at the fifth LaGrange Point, which is a position of stable equilibrium, equidistant from the earth and the moon. Since its formation, the society's goals have broadened and become more sophisticated. Now it promotes a strong U.S. space program, seeks to make the public aware of the importance of space development, and is in many ways the most active space organization in the country.

The society operates a phone tree, which can be activated at any time to make strong inputs to Congress and to the administration on key space-related issues. For example, in 1979 and 1980 it was the efforts of the L-5 Society that alerted people of this country to the dangers of signing the United Nations "Moon Treaty," which offered a threat to free-enterprise activities in space. More recently, it has spearheaded the Citizen's Advisory Council for National Space Policy, a group of scientists, political leaders, astronauts, business people, and engineers dedicated to preparing a comprehensive plan for U.S. space development through the end of this century.

There are more than sixty local chapters of the society, in this country and throughout the world, many with their own very active programs. These include lectures and space films, local legislative actions, speakers' bureaus, local newsletters, and meetings to pool information, plan space fairs, and pursue space-related ideas. Although the L-5 Society itself is not a political lobbying organization, there is a political action committee (Spacepac) that does lobby, and that shares the same goals as the society. Opportunities exist for L-5 members also to become active in Spacepac.

Since 1982, the society has sponsored an annual Space Development Conference, containing workshops with business, academic, government, and space-advocate participants. The conference offers a mixture of technical and general sessions, conducted in parallel programs.

The *L-5 News* is the society's national bimonthly newsletter, offering up-to-date news about technical, social, economic, and political matters relating to space. It is particularly sensitive to affairs in Washington, D.C., that may affect U.S. space development.

Annual membership categories and dues are as follows:

Student member	$15.00
Regular member	$25.00
Supporting member	$50.00
Contributing member	$100.00
Sponsoring member	$200.00
Lifetime Member:	$500.00

For additional information about the society, contact the international headquarters at 1060 East Elm, Tucson, Arizona 85719 (Telephone: 602-622-6351).

THE NATIONAL SPACE CLUB

Although the National Space Club's activities are in many cases focused on events in the Washington, D.C., area, its members include individuals and corporations nationwide.

It was formed in 1957 as the National Rocket Club, established to stimulate the exchange of ideas about rocketry and astronautics and to promote the recognition of United States achievements in space. It is a nonprofit corporation that addresses itself to "the broad historical, educational and business aspects of the missile and space fields."

More specifically, the activities and objectives are stated by the club to be:

- To promote United States space leadership through the media of conferences, the press, and other literary and educational means.
- To stimulate the advancement of civilian and military applications of rocketry and astronautics and related technologies for the benefit of all mankind.
- To bring together persons from the federal government, educational institutions, the press, and other walks of life, for the exchange of information on rocketry and astronautics, and through them to inform the public at large.
- To provide suitable recognition and do honor to individuals and organizations that have contributed to the advancement of rocketry and astronautics.

To accomplish these objectives, the following activities are sponsored by the club:

- Monthly luncheons where recognized leaders speak on timely subjects and issues.
- The annual Goddard Memorial Dinner to honor individuals and institutions for their recognized contributions or achievements.
- Scholarships and awards to outstanding students to stimulate youth interest in astronautics and related sciences.
- Special functions as dictated by specific situations and events.
- Periodic newsletters to membership announcing club activities and programs.

Membership is limited to U.S. citizens and corporations. Annual dues are as follows:

Individual member	$15.00
Corporate member	$500.00
Small business	$250.00

Additional information concerning club membership and activities may be obtained from:

> The National Space Club
> 1629 K Street, N.W.
> Suite 700
> Washington, D.C. 20006
> (Telephone: 202-296-4690)

THE NATIONAL SPACE INSTITUTE (NSI)

Founded in 1974 by the late Wernher von Braun, this is the oldest and the largest of the societies that serve a general membership interested in space. As described in the group's own brochures, "The National Space Institute is a privately supported educational and scientific organization dedicated to communicating to the general public and the nation's leaders the importance of space technology as a resource for solving many of the world's most pressing problems . . . [It] is the voice of a growing membership of individuals from all walks of life who share a fascination with unlocking the mysteries of space and a concern that the way to a better tomorrow is through a strong space program today."

The society's charter sets forth its purposes as:

- To promote United States space leadership through the media of conferences, the press, and other literary and education means without, however, carrying on propaganda or attempting to influence legislation.
- To stimulate the advancement of civilian and military applications of space and related technologies for the bene-

fit of mankind and, if necessary, the defense of the
United States against aggression.

- To bring together persons from the federal government,
  industry, educational institutions, the press, and other
  walks of life, for the exchange of information on space
  and through them to inform the public at large.
- From time to time, to provide suitable recognition and do
  honor to individuals and organizations that have contributed to the advancement of space.

The society emphasizes grass-roots support for space,
and its board of directors and board of governors draw
from well-known figures of the entertainment and political
fields, as well as from science and industry.

Membership benefits include:

- Free subscription to *Space World,* a magazine that covers
  general space news and offers timely articles on particular
  space missions. Included with *Space World* is *INSIght,*
  NSI's own newsletter providing membership information
  and facts concerning upcoming events in which NSI will
  be involved.
- Space workshops, in which particular space-related topics
  are worked through in detail by participants, NSI staff,
  and consultants.
- Discounts on books and other materials related to space.
- Opportunity to join NSI group events and regional meetings, to attend shuttle launches, and to tour NASA and
  other facilities.
- Representation in Washington, to communicate with
  Congress, with NASA, and with other government agencies.
- Twenty-four-hour access to recent space news through a
  telephone "hot line" in Washington, D.C.

NSI membership dues are thirty dollars per year. For additional information on the organization and its functions,
contact the national headquarters at:

National Space Institute
West Wing Suite 203
600 Maryland Avenue, S.W.
Washington, D.C. 20024
(Telephone: 202-484-1111)

THE PLANETARY SOCIETY

The society was formed in 1980 by Carl Sagan, Bruce Murray, and Louis Friedman. Its general objective is to encourage the exploration of the solar system and the search for extraterrestrial life. More specifically, the goals are:

1. To encourage a realistic continuing program of planetary exploration and the search for extraterrestrial life;
2. To serve as a focus for the many individuals and organizations who share those objectives;
3. To involve the public in the adventure of planetary exploration by helping to initiate new endeavors.

The society's first year coincided with the Saturn encounter of the Voyager spacecraft, and also with the appearance of Carl Sagan's hugely successful book and television series *Cosmos.* Helped by those events, the organization saw an immediate phenomenal growth. However, continued rapid development since that time has made it clear that public interest in deep-space exploration is more than a transitory phenomenon. Although its emphasis is on exploration and scientific analysis rather than on applications, the Planetary Society has become the world's largest space-interest group, with more than one hundred thousand members. Initially active in the United States, it now has members on every continent.

The society promotes a realistic and continuing program of planetary exploration, and encourages research and development programs in national and international space agencies for exploration of the planets and the search for extraterrestrial life. In addition to these activities, the society directly funds research that will "seed" new exploration efforts. Such projects include the development of a radio receiver to be used in the search for extraterrestrial intelligence, a contribution for a new optical detector to make possible the discovery of extrasolar planets, support for work leading to the discovery of new near-Earth asteroids, and the development of a program of study for future Mars exploration. Most recently, the society has become increasingly active in science education, especially the production of resources and science materials for use in schools.

The society publishes a bimonthly color magazine, *The Planetary Report.* This contains the latest information and results about solar-system exploration, together with articles about upcoming missions, significant historical achievements in planetary sciences, and general space news.

Other services to members include provision of book and picture materials, lecture series, specialist conferences, and large-scale national meetings (such as the Planetfest in Pasadena, that coincided with the Saturn fly-by of Voyager Two).

Membership in the society is open to all and is intended for the general public. Annual dues, which include a subscription to *The Planetary Report,* are fifteen dollars. For additional information about the society, including a list of books, posters, slide sets, and maps available, write to:

The Planetary Society
P.O. Box 91687
Pasadena, CA 91109
(Telephone: 213-793-5100)

SPACE STUDIES INSTITUTE (SSI)

This is a privately funded, nonprofit organization devoted to "developing the resources of space for the benefit of all humanity." Founded in 1977 by Dr. Gerard O'Neill, SSI conducts research in techniques that will permit space utilization and ultimate colonization.

Typical SSI projects include reaction chemistry experiments to separate lunar and asteroidal materials into metals, silicon, and oxygen; techniques to locate and track small asteroids that might be brought at low cost to orbits around the earth; development of electromagnetic launch devices that can be used to transfer materials from the moon, between orbiting stations, and from the asteroids; and design studies for large space colonies, spaceborne industrial plants, and power satellites.

The organization is supported by several categories of members, with dues as follows:

Sustaining Member	$15.00
Contributor	$25.00
Donor	$100.00
Sponsor	$200–500.00

The category of senior associate is composed of individuals who pledge funds on a five-year or longer basis. All dues are tax-deductible.

SSI publishes a quarterly newsletter on research developments, and also organizes biennial springtime Princeton/SSI Conferences on the subject of Space Manufacturing.

For additional information concerning SSI, contact:

Space Studies Institute
195 Nassau Street
P.O. Box 82
Princeton, NJ 08540
(Telephone: 609-921-0377)

SPACEWEEK

This is not so much a society as an event. The Spaceweek organization exists to put on an annual, nationwide campaign to publicize space-program activities. It is held each

July, from the sixteenth to the twenty-fourth (the dates of the 1969 Apollo 11 mission), and it features lectures, telescope star parties, exhibitions, concerts, business promotions, model rocket launches, signing of proclamations, and every form of space-related function. All programs are heavily oriented to youth participation, including school contests in the spring with prizes awarded during Spaceweek.

Spaceweek does not have a formal membership plan. Instead, it is organized on a city-by-city basis. In each city where the group is active, there is usually one person who acts as the central point for contact and coordination. For convenience in planning, this individual is termed the director of the city's activities. In large cities or regions, the director is chosen annually by space-program supporters in the area. In smaller towns, the director may be self-appointed—a chance to start your space-program involvement at the top. In practice, the sponsorship of some organization that can offer labor and perhaps financial assistance is highly desirable, and a network of individuals willing to put in time and effort is also necessary for success.

Advice and assistance to any Spaceweek group can be found from Spaceweek national headquarters, where a group of seasoned space supporters will provide promotional materials, ideas, and access to national organizations and media. A *Spaceweek Handbook* has been developed, also a registration kit, bumper stickers, T-shirts, posters, slides, and a list of helpful publications and useful information. Progress reports around the country, together with suggestions for fund-raising and area events, can be found in the bimonthly *Spaceweek Newsletter*.

Major Spaceweek activities in 1983 took place in Austin, Texas; Boise, Idaho; Boston, Massachusetts; Chicago, Illinois; Columbia, South Carolina; Houston, Texas; Lafayette, Indiana; Los Alamos, New Mexico; Sacramento, California; San Diego, California; southern Washington State;

St. Louis, Missouri; Toledo, Ohio; Tucson, Arizona; and Upland, South Carolina.

For more information on activities in your local area, write to:

> Spaceweek National Headquarters
> P.O. Box 58172
> Houston, Texas 77258
> (Telephone: 713-332-3804)

STUDENTS FOR THE EXPLORATION AND DEVELOPMENT OF SPACE (SEDS)

This organization is unusual among national space groups in that it is focused on student participation. Although other societies permit student members, in SEDS the student element is central. The society was organized and is run by students.

Founded in 1981, the group aims at international political activity, through the mechanism of Nongovernment Organization (NGO) status with the United Nations. It seeks to become one of the six hundred NGO members of ECOSOC, the United Nations' Committee for Economic and Social Affairs.

Two categories of membership are recognized: student (annual dues ten dollars) and nonstudent (annual dues twenty dollars).

The society has a quarterly newsletter. Headquarters are in Washington, D.C., and additional information may be obtained from SEDS, 800 21st Street, N.W., Box 24, Washington, D.C. 20052.

UNIVERSITIES SPACE RESEARCH ASSOCIATION (USRA)

This organization was formally created in 1969, but its origins go back to 1966, when the NASA administrator asked the help of the National Academy of Sciences in creating a national consortium to manage the Lunar Receiving Laboratory. An original group of forty-five universities, now increased to fifty-two, was formed. It was created for the purpose of:

(a) Constituting an entity by means of which universities and other research organizations may cooperate with one another, with the government of the United States, and with other organizations toward the development of knowledge associated with space science and technology; and

(b) Acquiring, planning, constructing, and operating laboratories and other facilities, and formulating policies therefor under contract with the government of the United States or otherwise for research, development, and education associated with space science and technology.

In more informal words, USRA is an organization that forms a bridge between NASA and other U.S. government space-oriented agencies on the one hand, and the university system on the other. As such, it is an important group that any student seeking a space career should know about. Among other activities, USRA supports atmospheric-processes research at NASA's Marshall Space Flight Center, operates the Institute for Computer Applications in Engineering (ICASE) at NASA's Langley Research Center, operates the Lunar and Planetary Institute in Houston, directs the Materials Processing in Space at Marshall Space Flight Center, and supports other programs at NASA's Lewis Research Center. The headquarters for USRA is in Columbia, Maryland, about twenty miles from Washington, D.C.

The member institutions who participate in USRA activities are as follows:

University of Alaska
Arizona State University
University of Arizona
Boston College
Brown University
University of California at Berkeley
University of California at Los Angeles
University of California at San Diego
Case Western Reserve University
University of Chicago

Cornell University
University of Denver
Georgetown University
Georgia Institute of Technology
Harvard University
University of Houston
University of Illinois at Urbana
Indiana University
Johns Hopkins University
University of Kansas
Lehigh University
Louisiana State University at Baton Rouge
University of Maryland at College Park
Massachusetts Institute of Technology
University of Michigan at Ann Arbor
University of Minnesota (Minneapolis)
University of New Hampshire
State University of New York (Buffalo)
State University of New York (Stony Brook)

New York University
Northwestern University
Ohio State University
Old Dominion University
Pennsylvania State University
University of Pittsburgh
Princeton University
Purdue University
Rensselaer Polytechnic Institute
Rice University
Rockefeller University
Stanford University
Texas A&M University
University of Texas (Austin)
University of Texas (Dallas)
University of Toronto
Virginia Polytechnic Institute
University of Virginia
University of Washington
Washington University (St. Louis)
College of William and Mary
University of Wisconsin (Madison)
Yale University

These universities are organized into nine regional groups. For this and other general information concerning USRA's structure and functions, contact:

Universities Space Research Organization
The American City Building, Suite 311
Columbia, Maryland 21044
(Telephone: 301-730-2656)

In addition to the national and international organizations already described, there are many regionally based, special interest, or smaller groups scattered around the United States. Many were formed in the past three years, which has seen a strong resurgence of general interest in space and space exploration. Here is a partial listing of some of the more active and well known of them.

Aerospace Analysts Society
National Aviation and
  Technology Corporation
630 Fifth Avenue
New York, NY 10020

Aerospace Department
  Chairman's Association
Department of Aerospace
  Engineering
Auburn University
Auburn, AL 36830

Aerospace Education
  Foundation, Inc.
1750 Pennsylvania Ave-
  nue, N.W.
Washington, D.C. 20006
(Telephone: 202-637-3370)

Aerospace Electrical Society
P.O. Box 248B
3 Village Station
Los Angeles, CA 90024

Air Force Association
1750 Pennsylvania Ave-
  nue, N.W.
Washington, D.C. 20006
(Telephone: 202-637-3300)

AMSAT: Radio Amateur
  Satellite Corporation
P.O. Box 27
700 Seventh Street, N.W.
Suite 224
Washington, D.C. 20044
(Telephones: 202-448-8649,
  301-589-6062)

Association of Lunar and
  Planetary Observers
P.O. Box 3AZ
University Park
Las Cruces, NM 88001

Astronomical League
4 Klopfer Street
Pittsburgh, PA 15209

Aviation/Space Writers
  Association
Cliffwood Road
Chester, NJ 07930
(Telephone: 201-879-5667)

California Space Institute
Mail Code A-030
University of California at San
  Diego
La Jolla, CA 92093
(Telephone: 714-452-4772)

Campaign For Space
P.O. Box 1526
Bainbridge, GA 31317
(Telephone: 912-246-6765)

Chicago Society for Space
  Studies
4N 186 Walter Drive
Addison, IL 60101
(Telephone: 312-529-1049)

Congressional Space Caucus
Representatives Daniel Akaka
  and Newt Gingrich, Co-
  Chairmen
1510 Longworth Office
  Building
Washington, D.C. 20515
(Telephone: 202-225-4906 and
  -4501)

Congressional Staff Space
  Group
322 House Annex #1
Washington, D.C. 20515
(Telephone: 202-226-2302)

Council of Defense and Space
  Industry Associations
1909 K Street, N.W.
Washington, D.C. 20006
(Telephone: 202-892-3970)

Foundation for Scientific
  Progress And Continual
  Exploration (SPACE)
616 FM 1960 West, Suite 601
Houston, TX 77090
(Telephone: 713-893-1332)

The GEOSAT Committee
153 Kearny Street, Suite 209
San Francisco, CA 94108
(Telephone: 415-981-6265)

The High Frontier, Inc.
1010 Vermont Ave., N.W.,
  Suite 1000
Washington, D.C. 20005
(Telephone: 202-737-4979)

Independent Space Research
  Group
P.O. Box 1246
Troy, NY 12180
(Telephone: 716-464-0125)

Institute for Security and
  Cooperation in Outer Space
201 Massachusetts Ave., N.E.
Suite 102-A
Washington, D.C. 20002
(Telephone: 202-547-3363/SOS
  alert: 202-547-3336)

Institute for the Social Science
  Study of Space
P.O. Box 922
Georgetown University
Washington, D.C. 20057
(No telephone listed)

Institute of Space and Security
  Studies
7720 Mary Cassatt Drive
Potomac, MD 20854
(Telephone: 301-983-1483)

International Alliance for
  Cooperation in Space
P.O. Box 5144
FDR Station
New York, NY 10150
(Telephone: 212-288-5704)

Maryland Space Futures
  Association
3112 Student Union Building
University of Maryland
College Park, MD 20742
(Telephone: 301-454-4234)

OASIS (Organization for the
Advancement of Space
Industrialization and
Settlement)
P.O. Box 704
Santa Monica, CA 90406
(Telephone: 213-374-1381—
message machine only)

Niagara University Space
Settlement Studies Project
Sociology Department,
Niagara University, N Y
14109
(Telephone: 716-285-1212,
extension 552)

Progressive Space Forum
1476 California Street, #9
San Francisco, CA 94109
(Telephone: 415-673-1079)

Royal Astronomical Society of
Canada
124 Merton Street
Toronto, Canada M4S 2Z2
(Telephone: 416-484-4960)

San Francisco Space Frontier
Society
2003 Lyon Street
San Francisco, CA 94115
(Telephone: 415-346-5421)

Space Cadets of America
256 South Robertson
Boulevard
Beverly Hills, CA 90211
(Telephone: 213-652-6452)

The Space Coalition
97 East Saddle River Road
Saddle River, NJ 07458
(Telephone: 201-934-1515)

Space Development
Foundation
2424 Pennsylvania Ave-
nue, N.W.
Suite 100
Washington, D.C. 20037
(Telephone: 301-474-5760)

The Space Foundation
114 Byrne Street
Houston, TX 77009
(Telephone: 713-864-4400)

Spaceweek, Inc.
P.O. Box 58172
Houston, TX 77258
(Telephone: 713-332-3804)

Sunsat Energy Council
P.O. Box 201
163 Main Street
Cold Spring, NY 10516
(Telephone: 914-265-3579)

United Futurist Association
P.O. Box 17059
San Diego, CA 92117
(Telephone: 714-272-4994)

United States Space Education
Association
746 Turnpike Road
Elizabethtown, PA 17022
(Telephone: 717-367-3265)

University of California Space
Working Group
Bechtel Engineering Center
University of California at
Berkeley
Berkeley, CA 94720
(No telephone listed)

Using Space for America
  (USA) Committee
1850 Columbia Pike, Suite 702
Arlington, VA 22204
(Telephone: 703-920-5670)

War Control Planners, Inc.
Box 19127
Washington, D.C. 20036

World Security Council
Suite 275-C, World Trade
  Center
San Francisco, CA 94111
(No telephone listed)

World Space Center
221 West Carillo Street
Santa Barbara, CA 93101
(Telephone: 805-965-7947)

World Space Foundation
P.O. Box Y
South Pasadena, CA 91030
(Telephone: 213-441-2630)

WRITE NOW!
P.O. Box 36851
Los Angeles, CA 90036
(Telephone: 213-386-1454)

5

Finding
a Job

In this chapter, we will concentrate on the central problem
of finding a job. Regardless of the amount and quality of
your training, the actual process of deciding where to look,
how to apply, and how best to impress prospective employers is still difficult. Before addressing specific space-related
markets, let us look at some prerequisites.

Selling the Product

In this case, you are the product. You are trying to sell
yourself and your talents. If you do not relish the idea that
you have suddenly become a salesman or saleswoman,
there are a few simple things to remember that can make
your task a lot easier. And if you do handle that first sale
correctly, you may never have to do it again in your whole
working life.

1) The first thing (and, if you do it wrong, the last thing) that a prospective employer is likely to know of you is through a job application. Your letter and résumé will probably arrive along with a dozen others. The first requirement is to pass the initial hurdle, that of interesting the person who reads your letter and résumé. Remember, all that person knows about you is contained in those pieces of paper you have sent along. Ninety percent of job applications fail to pass that initial screening and generate no more response than a polite rejection letter. There are several possible reasons for this.

First, when you apply for a job in government or industry, you should already have done some homework. You should know something about the organization—how big it is, how old it is, how many people it employs, and what it does. You want to know this last one in some detail—it is not enough to know, for example, that IBM "makes computers." The company also makes typewriters, copiers, and other office equipment; and if you are interested in space activities you are probably most concerned with their Federal Systems Division, which makes no general commercial computers and no office equipment. Ideally, you will know something about a specific project on which you would like to work, and refer to it in your résumé and letter of application. Applying for a job that does not exist wastes your time and the company's.

It's necessary to know what a possible employer does, because your application must try to answer a central question: namely, *Why should they be interested in you?* We know what you want from them, and it is important for you to keep the vision of what you want to create in your mind. You want a good, perhaps a specific, job. But before you are likely to be offered one, you need to know what *they* want.

That depends on their product lines, their contract backlogs, and their general adequacy of personnel. If you can find a way to suggest to an overworked manager that you can add something valuable or help solve one of his or her problems, instead of adding to them, the battle is almost won. He or she will take the internal action to get an offer out to you, as soon as possible. Don't entertain any illusions. You are not interesting to employers because of your wit, lovable nature, or pleasing personality—you are interesting as an element in solving problems they face. You may even be able to suggest and create a new job position if you see something that might complement a situation. The more you show knowledge of their problems and situation in your application, the better are your chances. The best technique of all is to try to find, through mutual acquaintances, someone who knows someone in the company or already works where you will be applying. Objective evaluation sounds desirable, but in the real world almost everything is done through people. A personal recommendation is worth a dozen good résumés.

2) The employer sees you initially only as you have presented yourself in your letter of application and your background résumé. No matter how talented and well trained you are, if this is not presented in your written material it is usually irrelevant—you will never get to the interview stage. For this reason, it is a bad idea to use a universal, "general purpose résumé," which can be copied and sent out blindly to a hundred different places. Don't ever mail a messily Xeroxed résumé that suggests that it has seen wide circulation and multiple copying machines. Nothing turns an employer off faster —unless perhaps it is a résumé of the "kitchen sink" variety, in which the writer has tried to describe everything that has happened to him since he was ten years old in the hope that a few items will have special significance to the reader.

Keep it short. Busy people do not have the spare time to wade through a dozen pages of unstructured background material. A good résumé is no more than one page long. It takes time and effort to do it, but every application should be accompanied by a custom-made résumé, one that has been developed to include only the things that make your particular background fit their particular needs. There is plenty of opportunity to amplify that brief description if you get to the point of an interview. The typical minimal information that should go into a résumé is summarized in Figure 5.1. Most important of all: Make sure your résumé contains your address and telephone number. People don't save envelopes, and accompanying letters have a nasty habit of becoming detached and disappearing.

Don't oversell what you have done. No employer will expect a tremendous string of accomplishments from someone who is applying for a junior position. And the principle is true at all levels. We have seen résumés from undergraduates that claimed more than the job description for the administrator of NASA or the Secretary of Defense. That's not plausible. If you talk of great experience and talent, it is more likely to produce suspicion and incredulity than a positive impression.

Résumé-writing is a special skill, and we have touched only on the most major points; we recommend that the reader examine a book that goes into the subject in detail—for example, the paperback text *Résumés That Get Jobs* (Arco Publishing Company, Inc., New York, 1977). Good résumés don't get jobs, of course; but bad ones often lose them.

Figure 5.1

THE BASIC RÉSUMÉ

Your name   Your address and phone number

JOB OBJECTIVE (in one short sentence)

Add one paragraph that describes you in relation to the position you seek.

EMPLOYMENT HISTORY

(in inverse chronological order, with your most recent job, if any, first)

Dates (start and end)	Name of organization, position held
Dates (start and end)	Name of organization, position held
Dates (start and end)	Name of organization, position held
Dates (start and end)	Name of organization, position held

(Don't list your reasons for leaving—save that for the interview)

Part-time employment, if appropriate to job applied for.

EDUCATION

(inverse chronological order)

Dates (start and end)	Institution	Degree/diploma/honors
Dates (start and end)	Institution	Degree/diploma/honors
Dates (start and end)	Institution	Degree/diploma/honors

Relevant additional education, courses, etc.

List any scholarships won, honors obtained, sports and other activities. Be brief, and restrict your other activities to ones relevant to the job.

PERSONAL DATA

Age, marital status, relevant societies and hobbies, willingness to travel or relocate, general health and physical description. Personal references.

PRESENT SALARY AND SALARY REQUIREMENT

(Don't include present salary if you are changing career paths and expect to take a pay cut to do it—unless you also want to state that fact explicitly in the résumé. Many people overprice themselves out of jobs they would have been happy to have taken at a lower figure.)

On any item where there is a choice as to what to include, select only those activities or facts that are relevant to the particular job you are seeking. Unnecessary detail *decreases* your chances of success.

3) If you pass the hurdle of initial application, you now face the scarier one of the employment interview. To many people, this is the worst part of job application. Rejection after an interview is always harder to take than rejection of a letter or a résumé, because it feels more personal.

   We can't make interviewing easy, but we can offer a few simple rules that make for interview success.

a) Basics: Get there on time, be polite, dress for the job, don't pick your nose or indulge in other socially unconventional behavior. This applies even if the job you are applying for is to be done in the middle of the night with no human contact.

b) If you don't know the answer to a question, say you don't. There's no disgrace in ignorance, but there is in lying—and as often as not you'll be found out. (At one of the authors' first interviews, the interviewer mentioned several times that the computer work included "Monte Carlo calculations" for nuclear-reactor calculations. On the third reference, the nervous applicant finally confessed that he didn't know what that meant. "I thought you didn't," said the interviewer. "When I came here, the same thing happened to me.")

c) Be as natural as you can be. Employers know that interviews are stressful, so they make allowances for shyness, and for a certain amount of babbling.

d) Listen closely to questions and comments, without interrupting. If you don't understand, ask for a clarification. A lot of what you hear is likely to be unfamiliar to you.

e) Take time to think a moment before you answer questions. And don't simply try to give the answer that you think the employer wants to hear. If you don't agree, politely say so—it may be an interview test.

f) Be very careful how you talk about previous jobs you have held, or the job you hold now. Unless you are asked, we suggest that you not bring up the subject at all. And if you do talk about it, make it as positive as possible. There is nothing to be gained from denigrating your past employers. The best reason for leaving a job is that you want the job you are now applying for more.

g) Don't hassle the interviewer by asking about your long-term financial or promotion prospects in the organization. Questions like this are virtually unanswerable. Your progress will depend on your usefulness, how you stack up against other good people hired, the general growth of the group, the state of the economy, and a number of other imponderables.

h) Be enthusiastic. Energy, initiative, and enthusiasm are more important in practice than educational background, age, or experience.

i) If the interview seemed a total disaster, don't be too upset. A positive attitude will help you find a position. Besides, it may not be your fault. Interviewers suffer stress, setbacks, overwork, and depression in their jobs—you may have caught one on a day when no applicant could have pleased. Take heart. Remember, Western Union turned away Alexander Graham Bell and his telephone; RCA, GE, and IBM turned down the basic Xerox patents; Parker Brothers came very close to rejecting the game of *Monopoly* when it was offered to them (because it "took too long to play"). And most employers today would show Albert Einstein the door.

Put the rejection behind you, and apply somewhere else. (*After* you hear that you have been turned down. Interviewers have been known to be abrupt and dismissive because they decided to hire you in the first two minutes, and they were running late for another appointment.)

4) What about personnel agencies? These are organizations that specialize in finding jobs for people, and people for employers. It sounds like the ideal solution, since they know the market situation better than you are likely to, and their whole livelihood depends on matching job skills with job needs.

We have mixed feelings on these. First, many survive by placing people. Even if you are a round peg and the job slot is square, they may do their best to squeeze you into it. Remember, it is your career and not theirs that is being manipulated. Second, you need to define clearly at the outset who will be paying their fees—you, or your future employer? Some "executive search" or personnel agencies will charge you a great deal of money to help to rewrite your résumé; but then they may prove of little use in assisting you to find a position.

If you do go to a personnel agency, make sure that the individuals you are dealing with have some real knowledge of the field, and a real interest in it, and in you. It may be that you will be much better served through the job information and job placement services provided by many of the professional societies and organizations described in the previous chapter.

The Changing Job Market

"I think there is a world market for about five computers."
—Thomas J. Watson, Sr., founder of IBM Corporation
(speaking in the late 1940's)

"The 1990's will differ from the 1970's as profoundly as the nineteenth century from the eighteenth."
—Clive Sinclair

Sinclair, designer of the first pocket-sized calculator and of the Timex-Sinclair microcomputer, was talking particularly about the job market and the use of people. A hundred years ago, three quarters of the American work force were employed in some aspect of agriculture and food production. Today, only one person in fifty works on primary food production—but far more than that are employed in the "value-added services" associated with food, from packaging to shipping to advertising, to supermarket checkout and restaurant service. Today, we often get the impression that as many people are employed in the diet business as in direct food production.

This may seem far removed from the subject of space careers, but it isn't. The proportion of this country's work force engaged in the primary production industries (food, raw materials, energy production) is shrinking steadily, while the proportion engaged in forms of service grows. At the same time, the traditional major industries of this country, such as steelmaking, automobiles, aircraft, and plastics have reached maturity and are being challenged more and more by production from other nations, particularly those countries where labor rates are lower.

All these changes point to a radical shift in job opportunities over the next twenty years. And the rate of change continues to increase. Some of the skills in demand today may have no value a decade from now, just as the butter-churning, weaving, and horseshoeing expertise of a century ago has no value in present markets.

Estimating where the markets will bloom and where they will shrink is a chancy business. It is even more difficult to estimate which particular companies will do well, since the skills and foresight of their managements play such a large part in their success. However, we see no end to increased opportunities in the following business fields:

- Computer applications—in every field, from graphics and animation in entertainment, to political forecasting, world security and monitoring systems, medical analysis, video games, spacecraft control, automobile engines, checkless bank accounts, food distribution, championship chess-playing, and a thousand others.

- Preventive medicine and high-technology health care. This will be both a spin-off from and a contributor to long-term space occupancy, with rapid growth of miniaturized implants for monitoring and controlling abnormal body conditions, and increased use of computer-controlled online patient-care systems.
- Entertainment and leisure-time industries. As automation takes over the production side of manufacturing, enforced idleness and more fun in jobs may become normal for much of the population. One outstanding point about space exploration is that it has high interest and entertainment value for many people. With a large enough viewing audience, projects such as the Voyager and Galileo exploration of Jupiter and Saturn justify themselves on interest value alone.
- Communications, of every form. We will see jobs in transmission and processing of electronic mail, direct-broadcast television, personalized newspapers (available in print and through terminals), international conferencing, and financial data. Instant debit and credit of bank accounts is almost here, as is instant general checking of consumer credit status. In the home, it is one of the major mysteries of the past twenty years that industry has been so slow to recognize the potential uses of the television screen for more than commercial viewing. The combined computer terminal, two-way picture data-transmission device (picturephone), and entertainment device is almost here. Its arrival will generate vast numbers of new jobs.
- Education. This will, like many other professions, have a large computer-assisted component. But the 1980's is the period when the children of the 1950's baby boom arrive in the school system in large numbers, to create a second

wave of school openings, teacher shortages, and children-based industries. This is a good time to create jobs that will enhance educational needs of students, families, and educators to prepare for accelerating changes.

- Mass-produced, personalized services. If this sounds like a contradiction in terms, note that it is already happening. It used to be that cars, telephones, shoes, games, or airline tickets came in just a few basic styles. Now, with production and service systems computerized, infinitely variable designs, styles, and routes from anyplace to anywhere else can be developed by the machines just as easily as the standard models or travel plans. Stock inventories will decrease—you only make what is actually ordered—but choice will increase.

- Dial-up and delivery services. The changes here will be forced by two factors, one social and one technological: There will be more old people (the average age of humans is steadily increasing and will continue to increase through the end of the century), who are less inclined to physical mobility but who want home services; and more powerful communications services will permit electronic mail, remote and real-time catalog ordering, and electronic payment of bills and checking of credit. The first electronic home-banking system is already in operation.

It takes no magic crystal ball to generate this list. In fact, unless something catastrophic happens, most of the predicted changes seem inevitable. Now, however, we have to relate them to jobs involving space, and that is a more difficult proposition. Before we begin, let us list another view of the most promising career areas for the 1980's. This is taken from *90 Most Promising Careers for the 80's,* by Anita Gates (Monarch Press, New York, 1982). A few other books on general career planning are listed at the end of this section.

Accountant	Agricultural Engineer
Actuary	Airplane pilot

Auto Body Repairer
Auto Salesperson

Bank Clerk
Bank Officer
Bank Teller
Biochemist
Biomedical Engineer
Business Machine Repairer

Cement Mason
Ceramic Engineer
Chef
City Manager
Claim Representative
Collection Worker
Computer Programmer
Computer Service Technician
Construction Inspector
Correction Officer

Dental Assistant
Dental Hygienist
Dental Laboratory Technician
Dentist
Dietician
Display Worker
Drywall Installer or Finisher

Economist
EEG Technologist/Technician
EKG Technician
Emergency Medical Technician
Engineering and Science
  Technician

FBI Agent

Flight Attendant
Floor Covering Installer
Floral Designer
Forestry Technician

Geologist
Geophysicist
Glazier
Guard

Health Services Administrator

Industrial Engineer

Landscape Architect
Lawyer
Licensed Practical Nurse
Life Scientist

Marketing Researcher
Medical Laboratory Worker
Medical Record Technician/
  Clerk
Metallurgical Engineer
Mining Engineer

Nursing Aide/Orderly/
  Attendant

Occupational Safety and
  Health Specialist
Occupational Therapist
Occupational Therapy
  Assistant
Operating Engineer
Optician
Optometric Assistant
Optometrist

Osteopathic Physician	Registered Nurse
	Respiratory Therapy Worker
Personnel Specialist	Retail Trade Sales Worker
Petroleum Engineer	
Pharmacist	Secretary
Physical Therapist	Sheet Metal Worker
Physician	Speech Pathologist
Podiatrist	Statistician
Police Officer	Systems Analyst
Psychologist	
Public Relations Specialist	TV/Radio Service Technician
Purchasing Agent	Travel Agent
	Underwriter
Radiologic Technologist	Urban and Regional Planner
Range Manager	
Real Estate Agent/Broker	Veterinarian
Receptionist	

Certainly not much sign of space careers, you might say, in that list, though there are many related jobs, such as systems analysts. Notice, however, the concentration on service work—very few of the positions listed are involved with *making* anything, and the vast majority are concerned with personalized maintenance and repair work (for both equipment and people).

Our own perceptions of the top jobs for the 1980's and 1990's are rather different from the ones just listed. It is clear from our list of growing areas that we believe there will be increased application of computers to jobs traditionally considered human territory. In fact, we think that *anything* reducible to logical rules or routine operation is fair game for computer take-over. This includes much of accounting, some levels of actuarial work, many manufacturing positions (better done by smart robots), and even the simpler forms of cleaning and maintenance. Accountants may be no fewer in number, but they will spend far less time on the strictly mechanical (and boring!) chores of bookkeeping.

The personalized or "handcraft" jobs, particularly those involving other humans or any form of aesthetic judgment, are safe. It will be a long time before computers take over hairstyling, medical practice, or landscape gardening. At the same time, jobs that call for creativity or design ingenuity will be increasingly needed for the foreseeable future. Computer programmers will still be on the list of "most wanted" staff, but higher-level computer languages and more user-friendly systems will increase the need for system designers and decrease that for simple coders.

Combining the space element with the growing job markets of the eighties and nineties leads us to suggest that you aim for the following top-ranked areas.

Systems analyst: What does a space systems analyst *do*? That's a very difficult question. The simplest answer is that he or she studies the way in which the individual components of complicated systems interact with each other, and tries to produce a design that is efficient, safe, inexpensive, and as simple as possible. Quite often the interviewer for a systems-analyst position will be unable to define the field, so don't be intimidated if you have trouble doing so. The nice thing about systems analysis is that it rarely requires academic qualifications in that specific field.

Computer specialist: Computers and automation are likely to be the magic words of the 1980's and 1990's, an absolute guaranteed boom area. These fields already play a large part in space development, and this will be more and more true as computers continue to decrease in size and weight.

Engineer: It is a sad fact that the United States produces fewer engineers as a proportion of graduating college classes than almost any other technically advanced nation. This is generally bad—what Robert Jastrow has referred to as a "time bomb ticking away in our society." On the other hand, it means tremendous job opportunities here for engineers. Rockwell International, for example, states explicitly that "the vast majority of positions available for new graduates within our company are for those graduates completing a degree in an engineering field."

Communications specialist: This is a broad field, ranging from hardware design for data transfer between computers and between people to explanation of the features of a new line of telephones or microwave ovens. There will be continued need in the space field for all types of communications expertise, from the most highly technical (design or write about a communications satellite) to the most general (explain and "sell" new space programs to the people, White House, and Congress).

Nurse or paramedic: We do not include "physician" here, since for a physician interested in space, the job opportunities are excellent, and an organization exists specifically to provide for them (see the Aerospace Medical Association description of Chapter 4). However, more and more medical work is now being performed by nurses and by paramedics. In the next decade, we believe that most of the jobs in flight preparation, astronaut testing, pilot testing, and postflight analysis will be done more and more by paramedics and nurses, as flights to and from space become routine and on a more and more frequent basis. Note, too, that while some medical training has already occurred via two-way, audiovisual satellite communication, other training methods will be needed for those who inevitably will work and reside off the planet. New kinds of jobs will be created. Prepare to apply old and new skills.

Information specialist: As space-related data bases proliferate, we see a rapidly increasing need for people who know how to access the information, ask the right questions of a computerized data file, and generally work with the many different forms in which information is available. The subject is not library science, nor is it computer programming. It lies somewhere between the two, and it covers everything from data entry (the least glamorous activity) to data-base design to distribution of data. Opportunities in the whole information-processing field will grow at a great rate for at least the next twenty years, and a good piece of the growth will be in space applications.

Teacher: Surprised to see this on a list of important space careers? We are tempted to say that it is the most important space career of all. Although there may be limits to how far we can go to persuade skeptics in their thirties or forties of the long-term importance of space, teachers can use their skills to advocate the need to participate in the process of developing space without weaponry for the benefit of all. But children are less fixed in their views and need teachers who have a holistic, world perspective, and who are excited about space. Be a good teacher, and your own enthusiasm for space development is bound to rub off on your students, thus helping prepare them for space careers. In addition to that, there will be many job opportunities—we see an increased need for teachers, particularly math and science teachers, all through the 1980's.

We see these as the best employment prospects. What about the effects of the higher proportion of old people, or the great increase in available leisure time to which we referred? Well, we certainly believe that both those phenomena will create numerous jobs; but we have trouble seeing how either area produces *space-related* positions. Perhaps some reader can point out to us what employment potential we are missing.

Suppose that your area of competence and interest isn't remotely like anything on the list, but you are desperately keen to find employment that somehow involves our space program. Then your space job probably has to be within a very large organization, even if your tastes would otherwise lead you to a small group (see the next section). The biggest aerospace companies, as well as NASA, the Defense Department, and other government agencies, have positions in almost every conceivable job field. There are openings for everything in the alphabet: accountants, analysts, architects, artists, and auditors; builders, comptrollers, drivers, engineers, firemen, and guards; purchasing agents, pilots, personnel officers, plumbers, and public-relations specialists—you name it, you can find it.

However, a plumber's job within NASA tends to resemble a plumber's job anywhere else. No surprise there. And NASA's comptrollers spend their time on finances, not rocket design. You won't find yourself suddenly asked to fly as backup on a shuttle mission, or be called on to give advice when a satellite malfunction places it in a wrong orbit. On the other hand, you will enjoy much of the ambience of space activities, ranging from access to lectures to a general feeling of being an integral, if invisible, part of space development.

If this won't satisfy you, then you have to be ready to change your job field. If you are an auto mechanic, you may choose to apply for a job building a satellite, thus applying your skills to a space project. Or you may have to go back to school in a different study area.

Making a List

This section is written for the person who has absolutely no idea how to get started, and particularly no idea what type of organization he or she would like to join. For people who already know what type of working environment they want, this section can be skipped. If you don't feel you have that knowledge, then you ought to sit down and decide on a few basics.

In what size organization would you be happiest? We will treat the world as though it were black and white—no shades of gray permitted. Companies will be "very big" or "very small." We assume that you will be in either a multi-billion-dollar organization, employing many thousands of people (which could certainly include the government), or one employing not more than a few hundred. Let us also note one exception: If you want to be a teacher, in space or on earth, then the Big Company/Little Company comparisons discussed below are quite irrelevant. We believe that teaching is a uniquely important job, and urge you to pursue it without worrying about our list.

Big company, small company. Which one do you prefer? The pros and cons of big and little are as follows; again, we are describing typical situations—there will be exceptions for every case we quote.

Big company	*Small company*
Usually good security, company will not go broke (but sometimes whole divisions may disappear).	Poor job security. Small companies have high failure rates.
Good fringe benefits such as health plans, stock purchase plans, retirement plans.	Few fringe benefits, and those very variable.
Little opportunity to see how your efforts affect the big picture.	You can often see the results of your own efforts, directly, on the company's development.
Salary increases usually set by general company policy.	Salary negotiable—how badly are you needed?
Rate of promotion set by company policy.	Fast promotion possible.
You can do well financially, but unless you make it all the way to the top (often a long, hard road) you will not become rich.	Chance to become very rich if company grows and you own stock in it. There are many examples of this in the consulting and services field, and will be more in this decade.

Fair chance you'll be bored at least some of the time (we are quoting from a fair-sized sample of people, though you may be one of the lucky ones whose job is always fascinating).	You may be stressed and overworked, but you probably will not be bored.
Opportunity to change fields without leaving the company, by changing divisions.	Little chance of changing fields without changing jobs (and remember, longevity benefits, such as increased annual vacation, do not move when you move).
Opportunity to be very specialized and still do well in pay and position.	In a small company, it is dangerous to be too specialized—the broader you are, the more likely you are to advance.
Employment opportunities for just about every skill.	Opportunity only for particular technical skills, plus marketing, management.

Unless you have some reason to do otherwise, we suggest that you assign equal weights to each point, one for your preference, zero for the other case (so the sum of scores for Big Company and Small Company should be ten, the number of comparisons listed above).

Some people may consider a single factor (for example, job security, or the chance to become very wealthy) as paramount. In this case, scoring the rest is irrelevant—your choice is clear. If you find that you score Big Company and Small Company about evenly, then you should look at the gray area of medium-sized companies, and see if you can find one that possesses the right combination of properties for your needs. Note also our general preference: if you can find a position in a company that is oriented more to *services* than *products,* we believe that the growth potential will be better.

Where Do You Look?

We are still on the basics. Now that you think you know what size and style organization would suit you, what is the easiest way to look for a job? If you happen to be in college, job counselors will be on hand. In addition, most of the large organizations with job openings will visit the campus some time during the school year. However, if you are not in school, there are four or five easy ways to find out about employment opportunities, locally or nationally.

In our preferred order, the best sources are:

1) The classified advertisement section of the local newspaper. Anyone who is looking hard for people is likely to run an ad there. If it involves space, it will usually say so, because that is considered a drawing card by most companies. You will often find out-of-town job openings advertised, with the dates that interviewers will be in your area.

   This is a case where you break our rule about a thorough investigation of the company before any face-to-face meeting—you may not have time before the interviewers are in town. Collect all the written material you can lay your hands on at the interview. In such cases, also ask who pays the relocation expense if you get a job offer (but do it in a low-key way).
2) Space-oriented societies and interest groups. We recommend that you join one of the organizations mentioned in Chapter 4. This will give you access to information about space and space jobs, and allow you to meet people in your area with the same interests.
3) Trade magazines. You will usually find these at the public library (if you don't, ask if they can get them for

you). They describe the general job market for your field of interest, and often they carry specific advertisements of job openings. If you don't know the names of the magazines you need, a library can again help you get started.

4) The local chamber of commerce. This can provide you with a list of companies in the area, and often with a good deal more. For example, the chamber of commerce often knows of recent large contract awards or plant expansion plans, plus other functions such as open days, science fairs, and recruiting sessions. If you are interested in some other city, write to the appropriate chamber of commerce and ask for a list of companies. There may be a charge for this, but it will be a small one.

5) Your friends and family. It is an old theory that you are only three links away from anyone else in the country— someone you know will know someone who knows someone who knows Mr. X. The same method applies to jobs. No one in your immediate circle of acquaintance may be knowledgeable about space jobs, but it's a good chance they know somebody who is. Follow the chain.

6) The Yellow Pages. This is scraping the bottom of the barrel as a method for seeking employment, but it will give you company names, addresses, and telephone numbers, and those you certainly need when you are collecting information.

Using all these sources, it makes sense to make your own reference file on companies, jobs, and people. The more general knowledge you have of space activities, the more impressive you can be in interviews.

Jobs in Industry

Although NASA commands the headlines, the vast majority of space-related jobs in the United States are to be found in industry. NASA and the Defense Department, today's big spenders on space, rely very much on contractor support to achieve their objectives.

Aerospace companies range in size and resources from the one-person operation working from a home or small office to multibillion-dollar conglomerates. There are tens of thousands of aerospace and aerospace-support companies, with some to be found in every state. A full list is obviously impossible, but we will give the names and addresses of most of the largest. We will also mention a few smaller, service-oriented groups, which generally produce no hardware but have good reputations and probable growth potential.

If you have read the section of this chapter called "Selling the Product," you will know that the last thing you should do is to fire off an optimistic résumé to a random assortment of these companies. If one of them particularly interests you, write a letter to their Public Relations Office. Request a copy of their annual report, plus any other descriptive material they may have available describing company activities.

In the case of the largest companies, they all contain numerous different divisions, only one or two of which may have anything to do with space—for example, the Singer Company has sales close to three billion dollars a year, but they produce things as diverse as sewing machines and furniture, in groups quite separate from their Space Division activities. This is not necessarily a disadvantage, though at first sight it may seem that way. It is usually easier to transfer inside a company, once you are there, than it is to get in in the first place. You may be able to enter an organization with your present qualifications, in some area unrelated to space, then after a while arrange an internal move to a space-oriented division. It's not easy, but if you can see no other way to get where you want to be, it is worth a try.

One warning. These different company groups often operate as effectively independent corporations, which certainly complicates hopping about from one part to the next. You need to find out how the organization is structured in your first explorations. Companies will usually transfer any specific inquiries you may have to the appropriate point within the general corporate structure. Allow a reasonable time (a month or more) for a response. If nothing happens within, say, six weeks, a polite follow-up letter or phone call is in order, repeating the request for information.

If you find a company that sounds promising, take the steps recommended in "Selling the Product." Find out what they do in as much detail as you can, and see if any of your friends knows somebody who works there. Don't consider a job application with the group until you have a good data base to work from.

A SAMPLING OF AEROSPACE COMPANIES

These are a mixture of large conglomerates, service companies, and specialized aerospace companies. The symbol ** by a name indicates that the company is largely or wholly services rather than hardware products.

Aerojet-General Corporation
10300 North Torrey Pines Road
La Jolla, CA 92037

Avco Everett Research
  Laboratory
2375 Revere Beach Parkway
Everett, MA 02149

Bell Aerospace/Textron
P.O. Box 1
Buffalo, NY 14240

The Bendix Corporation
Aerospace-Electronics
  Headquarters
1000 Wilson Boulevard
Arlington, VA 22209

The Boeing Company
P.O. Box 3707
Seattle, WA 98124

Burroughs Corporation
1 Burroughs Place
Detroit, MI 48232

Bunker Ramo
35 Nutmeg Drive
Trumbull, CT 06609

Calspan Corporation**
4455 Genesee Street
Buffalo, NY 14225

Computer Sciences Corp**
2100 E. Grande Avenue
El Segundo, CA 90245

Comsat
950 L'Enfant Plaza, S.W.
Washington, DC 20024

Corroon and Black**
Wall Street Plaza
New York, NY 10005

Control Data Corporation
Aerospace Division
P.O. Box 609
Bloomington, MN 55420

Curtiss-Wright
1 Passaic Street
Wood-Ridge, NJ 07075

Digital Equipment Corporation
129 Parker Street
Maynard, MA 01754

E-Systems, Inc.
P.O. Box 226030
Dallas, TX 75266

Environmental Research Institute
  of Michigan**
P.O. Box 8618
Ann Arbor, MI 48107

Fairchild Camera & Instrument
404 Ellis Street
Mountain View, CA 94042

Fairchild Industries, Inc.
20301 Century Boulevard
Germantown, MD 20874

Garrett Corporation
9851 Sepulveda Boulevard
Los Angeles, CA 90009

General Dynamics Convair
  Division
P.O. Box 80847
San Diego, CA 92138

General Electric
Space Systems Division
Valley Forge Space Center
P.O. Box 8555
Philadelphia, PA 19101

BF Goodrich Company
500 South Main Street
Akron, OH 44318

Goodyear Aerospace Corporation
1210 Massillon Road
Akron, OH 44315

Grumman Aerospace Corporation
Mail Stop C14-05
Bethpage, NY 11717

Hazeltine
500 Commack Road
Commack, NY 11725

Hercules Incorporated
910 Market Street
Wilmington, DE 19899

Hewlett-Packard Corporation
3000 Hanover Street
Palo Alto, CA 93404

Honeywell Inc.
Honeywell Plaza
Minneapolis, MN 55408

Hughes Aircraft
Corporate Headquarters
Culver City, CA 90230

IBM Federal Systems Division**
18100 Frederick Pike
Gaithersburg, MD 20706

International Telephone &
  Telegraph Corporation
320 Park Avenue
New York, NY 10022

Itek Corporation
10 Maguire Road
Lexington, MA 02173

Lear Siegler, Inc.
2850 Ocean Park Boulevard
P.O. Box 2158
Santa Monica, CA 90406

Litton Industries
360 North Crescent Drive
Beverly Hills, CA 90210

Lockheed Corporation
P.O. Box 551
Burbank, CA 91520

LTV
P.O. Box 5003
Dallas, TX

McDonnell Douglas
5301 Bolsa Avenue
Huntington Beach, CA 92647

Marsh & MacLennan**
955 L'Enfant Plaza North, S.W.
Washington, DC 20024

Martin Marietta Corporation
6801 Rockledge Drive
Bethesda, MD 20034

Mitre Corporation**
Burlington Road
Bedford, MA 07130

Northrop Corporation
1800 Century Park East
Century City
Los Angeles, CA 90067

Perkin-Elmer Corporation
Main Avenue
Norwalk, CT 06852

Planning Research Corporation**
1500 Planning Research
    Corporation
McLean, VA 22102

Raytheon Company
141 Spring Street
Lexington, MA 02173

RCA Astro-Electronics
P.O. Box 800
Princeton, NJ 08540

Rockwell International
    Corporation
600 Grant Street
Pittsburgh, PA 15219

Satellite Business Systems
8283 Greensboro Drive
McLean, VA 22102

Scientific-Atlanta
P.O. Box 105600
Atlanta, GA 30348

The Singer Company
8 Stamford Forum
P.O. Box 10151
Stamford, CT 06984

Science Applications, Inc.**
P.O. Box 2351
La Jolla, CA 92038

Systems Development
    Corporation**
2500 Colorado Avenue
Santa Monica, CA 90406

Space Communications Company
1300 Quince Orchard Boulevard
Gaithersburg, MD 20878

Sperry Corporation
1290 Avenue of the Americas
New York, NY 10104

SRI International**
333 Ravenswood Avenue
Menlo Park, CA 94025

TASC**
1 Jacob Way
Reading, MA 01867

Teledyne
1901 Avenue of the Stars
Los Angeles, CA 90067

Texas Instruments
P.O. Box 225474
Dallas, TX 75265

Thiokol Corporation
110 North Wacker Drive
Chicago, IL 60606

Tracor
6500 Tracor Lane
Austin, TX 78721

TRW
7600 Colshire Drive
McLean, VA 22102

United Space Boosters, Inc.
P.O. Box 1626
Huntsville, AL 35807

United Technologies Corporation
1 Financial Plaza
Hartford, CT 06101

Sperry Univac
Aerospace Division
8008 West Park Drive
McLean, VA 22102

Western Electric Company
222 Broadway
New York, NY 10038

Western Union
1 Lake Street
Upper Saddle River, NJ 07458

Westinghouse Defense and
Electronics Systems Center
Baltimore-Washington
International Airport
P.O. Box 746
Baltimore, MD 21203

Which one of these companies, if any, is best for you? There is absolutely no way we can answer that question. You will have to go through the process of evaluating your own preferences, bargaining position, knowledge and aptitude, while at the same time learning all you can of companies that interest you. It may be a long process, but it can't be skipped if you want the right job, rather than just any job.

Jobs in Space Command

As we pointed out in Chapter 2, the fastest growing area for jobs involving space is that funded by the Defense Department. Most of the openings will continue to be with aerospace companies. In addition, however, there have been recent changes in *direct* employment opportunities within the Defense Department.

The most significant change occurred in June 1982, with the creation by the air force of a Space Command (SPACECOM). This is the first step in the probable consolidation of all air-force space activities, and possibly all U.S. military space activities, into a single unified multiservice command. The main points about SPACECOM are as follows·

1) Space Command began operations on September 1, 1982.
2) Headquarters is in Colorado Springs, Colorado.

3) The initial structure will be built around an existing staff of the Aerospace Defense Center.

4) A number of existing organizations will be reassigned to SPACECOM. These include parts of the Strategic Air Command involved in space surveillance and missile early warning systems (including the Defense Meteorological Satellite Program), Peterson Air Force Base, Clear Air Force Base, Thule and Sondrestrom Air Force Base.

5) The Consolidated Space Operations Center in Colorado Springs will transfer from Air Force Systems Command to SPACECOM. When fully operational, it will become known as the Swigert Space Center, in memory of the late astronaut Jack Swigert.

6) All military space-shuttle activities will become the responsibility of SPACECOM.

7) The air force is now creating a Career Progression Guide for the new command, providing career planning for officers and enlisted people who are interested in a space-operations career. Officer specialties include space-operations analysts, space-operations officers, manned space-flight operations officers, satellite-operations officers, astronauts, space-operations staff officers, and space-operations directors. Since the program is in its start-up phase, we will undoubtedly see more positions created in the next few years.

8) Air force planning staff have emphasized the need for suitable personnel to carry out the duties of the Space Command. There are shortages of trained technical personnel, and there is currently only a small base from which to draw future Space Command leaders. Present plans call for a *tenfold* increase in Space Command personnel by 1986.

This rapid growth and corresponding shortage of personnel means difficulties for the air force—but their problem may be to your advantage. Space Command's needs offer you excellent bargaining power if you are of the right age and have the right qualifications. A science degree is preferred for officer positions, but since large-scale retraining of existing personnel will be necessary, the academic qualifications requirement may prove flexible. Some of the required academic background may also be obtainable when already serving in the military, and there are good opportunities graduate studies (some include pay) within the Space Command's career plan. Air force planning staff suggest that there is no better career opportunity within the Defense Department than those that exist today in Space Command.

The long-term possible ramifications of a separate Space Command, which could finally be an entity with a strong separate identity comparable to that of the army, navy, or air force, are huge. However, the present change should not be misinterpreted as a move to an immediate large manned space presence (though supporting this would certainly improve your chances for a career in space), or necessarily to large spaceborne weapons systems (which might jeopardize chances for long-term space, or any other, career).

Many of the traditional warning, interdiction, and reconnaissance systems will certainly have an increased space component in the coming years, providing space-career opportunities in security areas. For example, the ground-based early warning systems can be replaced by more effective space systems, giving much more notice of possible attack, and the satellites of the Global Positioning System allow rapid and accurate locations to be determined anywhere in the world. But these are unmanned space systems. The role of humans in space for military operations seems much less clear. The Defense Department will continue to be a large user of the Space Shuttle, but it has not to date been a strong proponent of a permanent space station. Perhaps the Space Command will lead a way toward developing a joint international space station and a satellite-security-monitoring system. Just imagine the career possibilities!

We have given no more than the briefest summary here, for two reasons. First, rapid changes are still occurring. Second, detailed information is best obtained from the Defense Department directly, or if you are in college from an ROTC. For more information, get in touch with:

> Directorate of Space
> Space Plans Division
> United States Air Force
> AF/XOSX
> Washington, D.C. 20330

Retrofitting: The Problem of a Second Career

The person who seeks a career involving space after many years of work in some other field faces tough problems with no easy solutions. First, there is a reluctance on the part of some employers to hire anyone who cannot make twenty years of possible contribution to an organization. Second, there is a preference for energetic young people who see themselves one day as presidents and directors. Third, and most subtle, some people find themselves very uncomfortable interviewing a person who is both older and in many ways better qualified (albeit in a different field) than they are.

We can do nothing to change their reluctance and possible insecurities, but we would like to suggest several ways in which you can turn your age and previous experience to your advantage.

If you ask an employer what qualities they seek above all else in their employees, they are unlikely to put brilliance at the top. A manager has fewest headaches when the people working for him or her are reliable, responsible, and can handle crises. As a rule, older people are better than young ones in all three areas—or at least seem that way (it's not that you become less crazy after forty; it's that you can often hide it better).

If you have worked in another field for many years, make sure that you point out to any possible new employer these main points:

1) You held a job and were continuously employed for a long time. This alone is evidence of steadiness and reliability.

2) You are looking for another steady, interesting job, but you would like this one to involve space, which is one of your long-term interests. To get that job, you know that you may have to make certain sacrifices— maybe financial, maybe in the status of the job you will hold.

3) In terms of your experience, you may be overqualified for the job available, but if you are not worried by this, the employer should not be. Make it clear that you are not worried—if you are overqualified and good, promotion opportunities usually create themselves.

4) List the qualities that your previous employment allows you to bring to a new job on the first day. Did you work for years in a high-pressure environment? Do you have experience handling personnel problems? Do you have sales experience? Management experience? Accounting experience? Shipping, travel-arrangements, or interviewing experience? Did you have to arrive at an exact time for years and begin work at once, with no scope for hangovers or all-night parties? If you have this background, make sure your résumé touches on it.

These are possible areas where you have a head start over young competitors. It makes sense to emphasize all of them (but again, in a low-key way; don't beat the drum loudly about these background qualifications).

Above all else, remember that you are coming in with at

least one strike against you. If you are looking for a second career, with a change of career path, it is essential that you remember that strike when you write your résumé or come for your interview. You must be not merely as well prepared as your younger competition—you must be better prepared. And with all that, it will not be easy.

State Aviation and Aerospace Departments

Each state has its own aviation department. Although they concern themselves with aircraft and aeronautics rather than space, and often have very limited resources, they usually know a great deal about what is happening locally. Contact them for information on air shows, science fairs, exhibits, or any local aviation or aerospace functions. Many space-oriented functions are held at and near airports and involve the state aeronautics groups.

Alabama	Department of Aeronautics Room 627 State Highway Building 11S Union Street Montgomery, AL 36130
Alaska	Department of Transportation & Public Utilities Pouch 6900 Anchorage, AK 99502
Arizona	Department of Transportation Room 118 205 South 17th Avenue Phoenix, AZ 85007
Arkansas	Division of Aeronautics Old Terminal Building Adams Field Little Rock, AK 72202
California	Transportation Commission P.O. Box 1499 Sacramento, CA 95807

Colorado	Department of Highways Aviation Transportation Section Hangar A Stapledon Field 8895 Montview Boulevard Denver, CO 80220
Connecticut	Department of Transportation- Aeronautics Bureau Drawer A Wethersfield, CT 06109
Delaware	Transportation Authority-Aeronautics Section P.O. Box 778 Dover, DE 19901
District of Columbia	Metropolitan Washington Airports Hangar Nine Washington National Airport Washington, D.C. 20001
Florida	Department of Transportation Florida Aviation Bureau 605 Suwannee Street Tallahassee, FL 32304
Georgia	Department of Transportation Bureau of Aeronautics 5025 New Peachtree Road, N.E. Chamblee, GA 30341
Hawaii	State Department of Transportation Air Transportation Facilities Division Honolulu, HI 96819
Idaho	Division of Aeronautics and Public Transportation 3483 Rockenbacker Street Boise, ID 83705
Illinois	Department of Transportation Aeronautics Division N. Walnut Street Springfield, IL 62706
Indiana	Aeronautics Commission 100 N. Senate Avenue Indianapolis, IN 46204

Iowa	Department of Transportation Aeronautics Division Municipal Airport Des Moines, IA 50321
Kansas	Department of Transportation Aviation Division State Office Building Topeka, KS 66612
Kentucky	Department of Transportation Aeronautics Division 419 Ann Street Frankfort, KY 40601
Louisiana	Department of Transportation & Development Aviation Office P.O. Box 4424 Baton Rouge, LA 70804
Maine	Department of Transportation Bureau of Aeronautics Augusta State Airport Augusta, ME 04330
Maryland	Department of Transportation State Aviation Administration P.O. Box 8766 Baltimore-Washington International Airport, MD 21240
Massachusetts	Aeronautics Commission Boston-Logan Airport East Boston, MA 02028
Michigan	Department of State Highways & Transportation Aeronautics Commission Capital City Airport Lansing, MI 48906
Minnesota	Department of Transportation Aeronautics Division Room 413, Transportation Building St. Paul, MN 55155
Mississippi	Aeronautics Commission P.O. Box 5 Jackson, MS 39205

Missouri	Department of Transportation Aviation Section P.O. Box 1250 Jefferson City, MO 65102
Montana	Board of Aeronautics P.O. Box 5178 Helena, MT 59601
Nebraska	Department of Aeronautics P.O. Box 82088 Lincoln, NE 68501
Nevada	Public Service Commission Capital Complex 505 East King Street Carson City, NV 89710
New Hampshire	Aeronautics Commission Municipal Airport Concord, NH 03301
New Jersey	Department of Transportation Division of Aeronautics 1035 Parkway Avenue Trenton, NJ 08625
New Mexico	Aviation Department P.O. Box 579 Santa Fe, NM 87503
New York	State Department of Transportation Airport Development Section 1220 Washington Avenue Albany, NY 12232
North Carolina	Department of Transportation Aeronautics Division P.O. Box 25201 Raleigh, NC 27611
North Dakota	Aeronautics Commission P.O. Box U Bismarck, ND 58505
Ohio	Department of Transportation Aviation Division University Airport 2829 W. Granville Road Worthington, OH 43085

Oklahoma

Aeronautics Commission
424 United Founders Tower Building
Oklahoma City, OK 73112

Oregon

State Department of Transportation
Aeronautics Division
3040 25th Street, S.E.
Salem, OR 97310

Pennsylvania

Department of Transportation
Aviation Bureau
Capital City Airport
New Cumberland, PA 17070

Rhode Island

Department of Transportation
Airports Division
T. F. Green State Airport
Warwick, RI 02886

South Carolina

Aeronautics Commission
P.O. Box 1769
Columbia, SC 29202

South Dakota

Department of Transportation Building
Pierre, SD 57501

Tennessee

Department of Transportation
Aeronautics Bureau
P.O. Box 17326
Nashville, TN 37217

Texas

Aeronautics Commission
P.O. Box 12607
Austin, TX 78711

Utah

Department of Transportation
Aeronautical Operations Division
135 N. 2400 W.
Salt Lake City, UT 84116

Vermont

Agency of Transportation, Aeronautics
    Department
133 State Street
Montpelier, VT 05602

Virginia

State Corporation Commission
Aeronautics Division
P.O. Box 7716
Richmond, VA 23231

Washington	State Department of Transportation Aeronautics Division Boeing Field 8600 Perimeter Road Seattle, WA 98108
West Virginia	State Aeronautics Commission Kanawha Airport Charleston, WV 25311
Wisconsin	Department of Transportation Aeronautics Division P.O. Box 7914 Madison, WI 53707
Wyoming	Aeronautics Commission Cheyenne, WY 82002

Books on General Career Planning

Resumes That Get Jobs, by Jean Reed, editor. New York: ARCO Publishing, 1977. (As we already remarked, even good résumés don't get you a job; they save you from losing one, and that's important.)

Essays on Career Education, by L. McClure and C. Buan, editors. Northwest Regional Educational Laboratory, 1973.

The Career Game, by C. G. Moore. National Institute of Career Planning, 1976. (A revealing title; a career is not a game, but getting a job may involve elements of gamesmanship, particularly if you happen to be interviewed by somebody who is playing their own game.)

International Jobs, by Eric Kocher. New York: Addison-Wesley Publishing Company, 1979.

90 Most Promising Careers for the 80's, by Anita Gates. New York: Monarch Press, 1982.

What Color Is Your Parachute?, by Richard N. Bolles. Berkeley, Calif.: Ten Speed Press, 1983.

Emerging Careers: New Occupations for the Year 2000 and Beyond, by S. Norman Feingold and Norma Reno Miller. Garrett Park, Md.: Garrett Park Press, 1983.

Careers Tomorrow: The Outlook for Work in a Changing World: Selections from *The Futurist,* edited by Edward Cornish. Bethesda, Md.: World Future Society, 1983.

Women in Space

"Nothing can be more absurd than the practice which prevails in our country of men and women not following the same pursuits with all their strength and with one mind, for thus the state, instead of being whole, is reduced to a half."
—Plato, Fourth Century B.C.

"We seek only a place in our nation's space program without discrimination. We ask as citizens of this nation to be allowed to participate with seriousness and sincerity in the making of history now, as women have in the past. There were women on the *Mayflower* and on the first wagon trains West, working alongside the men to forge new trails to new vistas. We ask that opportunity in the pioneering of space."
—from the 1962 congressional testimony of Geraldyn (Jerrie) Cobb, first woman to pass the U.S. astronaut-testing program.

Why Women?

Twenty years ago, this chapter would have been meaningless. Twenty years from now, it will perhaps be unnecessary. Today, however, the opportunities and obstacles facing women who wish to participate in the U.S. space program deserve to be addressed separately.

The idea of a "Women in Space" chapter—even a short one—raises a whole series of questions: Why women? Why do we devote a chapter to them, and not to blacks, Hispanics, American Indians, or many other minority groups? Is a woman working on space projects really treated any differently from a man?

These are fair questions. There are certainly inequities in opportunity for minority groups, based on educational defects, native language, cultural differences, and simple prejudice. We do not deny the importance of resolving these inequities; but we also feel that a discussion of women in space has a peculiar relevance today. We are at a threshold of increased women's participation in peace and in space matters; and we are also, coincidentally, at a threshold of increased weapons deployment in space. If space exploration is ever to achieve the potential that many of us see for it—as the "moral equivalent of war" that the philosopher William James sought at the end of the nineteenth century but that subsequent generations have failed to find—then women's involvement may be crucial.

Are there actually jobs that women are still excluded from—or are there others where they have a definite employment advantage? A discussion of the role that women play in the space program can take several different possible approaches. One, seen often in magazine articles, points to the minimal involvement of women in the past and deplores it. Another discusses only the careers of particular outstanding women. A third looks for any unique advantages that a woman might have in this field and suggests ways that she might be able to explore them.

Our subject in this book is space careers, not the general and complicated struggle for sexual equality. We argue that if there are factors that offer special advantages to women, it is our job to point them out. It is also our job to encourage women to enter this field which has been dominated by men, and to contribute new approaches (especially related to the peaceful development of space). And if there are pitfalls, or fields, where a woman still begins with two strikes against her, we want to mention them.

Slow Progress

History will record that the first woman in space was a Russian, Valentina Tereshkova, launched to orbit in June 1963. The first woman could easily have been an American. That she was not is a curious comment on the perceptions of the leaders of both countries in the early 1960's.

Tereshkova was a skydiver and textile worker, not a trained cosmonaut. She made one three-day trip to space, then dropped out of the Soviet flight programs to make numerous public appearances and lectures and to become a popular Soviet heroine. Nikita Khrushchev, the Soviet leader at the time, had correctly estimated the world's reaction to the flight of a woman cosmonaut. It was a triumph of favorable publicity, and Tereshkova performed her postflight task as ambassador and publicist for the Soviet program to perfection. This flight proved women, trained or untrained, can be space travelers. But her flight did not mark the entry of women as a regular and ongoing part of the Soviet program. No other Soviet woman went to space until 1981.

In the United States, the attitude of NASA and the White House seemed exactly opposite to that of the Soviet Union. The Russian, Tereshkova, was arguably underqualified for her flight. By contrast, long before that flight was made, over a dozen American women had passed the Mercury Program's astronaut tests and were itching for a chance to go into space. Jerrie Cobb, the holder of air-speed and altitude records, and a highly experienced pilot with over seven thousand flying hours (the average for selected male astronauts was only twenty-five hundred), had passed the Mercury tests with extraordinarily high scores back in February 1960. By the end of 1961, another twelve women had also passed those tests. But by then the bureaucratic decision had already been made and the door was closed: There would be no female astronauts in the Mercury or Gemini programs.

How was the decision justified? Well, there was the official logic, that NASA required any astronaut to be a jet-aircraft test pilot. This is, superficially, a reasonable requirement. But since the test-pilot schools were operated solely by the armed forces, and since at that time there were no women pilots permitted in the military services, any woman would-be astronaut was neatly trapped by the system. She could be in the military, or she could be a test pilot—but she could not attend the military jet-test-pilot school. And that was required to be an astronaut.

The true reason for the rejection of an outstanding group of qualified women candidates, however, ran much deeper. It reflected the mores of America in the late 1950's and early 1960's, within which going to space was not "right" for a woman. The view was held somewhat by the general public, and more strongly by the U.S. Defense Department. It is captured completely by an answer given by John Glenn during congressional testimony in July 1962:

Question: "How would you defend what is apparently a built-in discrimination against women in the space program, since jet-test experience is denied to them?"

Colonel John Glenn: "The men go off and fight the wars and fly the airplanes, and come back and help design and build and test them. The fact that women are not in this field is a fact of our social order."

Things *do* change. Today's Senator Glenn, who had fewer flying hours than Jerrie Cobb, would never dream of making a statement like that, on or off the record. And today, more than twenty years later, we finally have women astronauts. But anyone who reads Jerrie Cobb's autobiography cannot fail to feel the frustration and heartbreak behind her uncomplaining recitation of events of 1961–62, in which her superior qualifications meant nothing to the official decision process.

And the Defense Department, which controlled the input stream of potential astronauts in the early sixties, still has far to go in making sexual equality a matter of practice and not merely of law. It is perhaps the least progressive of all U.S. government organizations in opening up job opportunities to women. Here, in 1983, are the words of Mary Evelyn Blagg Huey, who is president of Texas Woman's University and also heads the Defense Advisory Committee on Women in the Services:

"The closing of military occupations impacts negatively on career development for women, making their advancement difficult if not impossible. In addition to the questionable legality of such direct consequences, the 'domino effect' poses concerns for morale, enlistments and the continued success of the all-volunteer Army."

She points out that the combat-exclusion policy for women automatically prevents their consideration for certain types of career advancement. This is very much the astronaut-selection trap of 1961, updated for the 1980's.

We do not deny that there has been progress since the Mercury Program. But a woman who wishes to develop a space career must still face certain hurdles that a man never sees. And those hurdles are perhaps higher in aerospace than elsewhere.

Working Rules

In addition to a handful of astronauts, women now work at all technical and most managerial levels in aerospace, government, and industry. But space, and particularly the aerospace industry, is one of the toughest fields for a woman to seek success in, probably because of its traditional ties to the strongly male-dominated military programs. Affirmative-action plans, required of companies doing government contract work, have gradually forced the employment of more women in engineering, design, and management. But this has also produced a male backlash of its own. In an effort to fill quotas, many aerospace companies have found they must pay *higher* salaries to women engineers than to men with the same qualifications. The result is considerable resentment by the latter of their women colleagues, regardless of their talents.

The experience of technically trained women, working in what have traditionally been male-dominated fields, is the source of this general advice to women entering the aerospace field:

- Be prepared to offer repeated and continuing proof of your competence. Women are still suspect to many men in the "hard" sciences of mathematics, physics, chemistry, and engineering.
- Beware of overprotectiveness, and assume responsibilities you know you can handle. Some men will insist that a woman do better than a man before she is respected, others will overprotect by excluding women from the hardest or high-pressure jobs. The reasons for doing this are varied, ranging from "this job is too dirty to give to a woman" to "our clients in the Middle East (or Africa, or Japan, or South America!) will not deal seriously with a woman representative." This exclusion from the toughest and dirtiest jobs not only limits career progress—it also produces resentment.
- Beware of discrimination from other women, and talk to your women colleagues about this. Although there are more and more exceptions, it is not uncommon for a woman manager or technician to find that she receives

less support from the secretarial pool than does the average man. In our experience, the professional woman is sometimes accused of a "superior attitude" by women support staff, and to find her work comes last on the priority list. Some women resent the success of a woman engineer or manager. And at the deepest and usually unspoken level, the male manager/female secretary relationship often reflects a general male-female sexual awareness and stimulation that may be relished by both parties.

• Don't stay too long in one job. Working your way to the top within a company may be more difficult for a woman, and after a few years in one place it is time to reevaluate and perhaps seek a move.

• Pay special attention to your relationship with other staff members, particularly when it comes to displaying emotion or ambition, or exerting pressure to force results. A man who has a fit of temper at delays or errors, or presses aggressively for rapid results, can often get away with it by saying it was done for the good of the project. A woman behaving in the same way is apt to be accused of "female instability" on the one hand, or "bitchiness" on the other (though seldom to her face).

• If you do encounter hostility or bias, try to handle it for yourself, if possible in a humorous way. A good deal of initial male uneasiness will evaporate as soon as it is clear that you are competent, not threatening his position, and not looking for any special treatment because you happen to be a woman. But speak up at once if you suffer real sexual harassment, or if you are being shielded (for whatever reason) from the most difficult or challenging assignments.

• Be careful if you decide to use the fact that you are a female to seek special favors or privileges. You may get them, but you may suffer a long-term loss of status and respect that more than cancels the apparent gain. We suggest you maintain a position of wanting to be respected for your technical or managerial merit alone while being respected as a woman.

• Be sensitive to the dangers of the "quota" system. In such cases, you may be invited to join an organization or professional group principally because you are female, and some quota of women employees must be filled to satisfy government requirements or company policy. The system sounds helpful to women's job needs, but there is a risk that the final effect may decrease the credibility and professional reputation of all the group members selected on such a basis. Nevertheless, if you get into a position this way, you've got the job . . . so do it.

Reference Works

There are hundreds of books designed to help women find jobs. We list only the few that seem relevant to a space career, plus a few others that are peripheral to a career search.

Woman Into Space, by Jerrie Cobb. Englewood Cliffs, N.J.: Prentice-Hall, 1963. (We recommend this book, now out of print but available through libraries, to give an idea just how difficult it was, twenty years ago, for even an outstanding woman to receive unbiased treatment by the U.S. institutions of the day. Given such treatment, Jerrie Cobb would probably have been the world's first woman astronaut).

Women's Networks, by Carol Kleiman. New York: Harper and Row, 1980.

Back to Business: A Woman's Guide to Reentering the Job Market, by Lucia Mouat. New York: Signet/New American Library, 1980.

Where the Jobs Are: An Annotated Selected Bibliography, by the Business and Professional Women's Foundation, 2012 Massachusetts Avenue, N.W., Washington, D.C., 1979.

Directory of Career Resources for Women, Ready Reference Press, P.O. Box 5169, Santa Monica, Calif. 90405.

Women and the Future, by Janet Zollinger Giele. New York: Macmillan, The Free Press, 1978.

Women and Work, by Carol Hyatt. New York: Warner Books, 1980.

Getting Ahead: A Women's Guide to Career Success, by S. Norman Feingold and Avis Nicholson. Washington, D.C.: Acropolis Books, 1983.

The Black Woman's Career Guide, by Beatryce Nivens. New York: Anchor Books/Doubleday, 1982.

Women in the Military, by Major General Jeanne Holm. Novato, Calif.: Presidio Press, 1982.

Information from NASA

NASA now has representatives of the Federal Women's Program in most field centers, to inform you about women's job opportunities and to provide general information. Contact the following individuals, or write to the Federal Women's Program at the appropriate center (field-center addresses are given in Chapter 2).

NASA Headquarters: Mary Jackson (acting manager), 202-755-3714.
Ames Research Center: Wanda Hunter, 415-965-5778.
Goddard Space Flight Center: Mary Jo Sharp, 301-344-5715.
Marshall Space Flight Center: Betty Aldridge, 205-453-4202.
Kennedy Space Center: Jay Diggs (acting manager), 305-867-2307.
Langley Research Center: Mary Jackson, 804-827-3487.
Johnson Space Center: Estella Gillette, 713-483-4831.
Lewis Research Center: Monica Crespo, 216-483-6957.

National Space Technology Laboratories: Kris Butera,
601-688-1912.

For advice on opportunities or problems for women in
the space program, the general women's organizations at
the national level seem to have little useful information.
We suggest that you contact one of the following specialist
groups:

The Hypatia Cluster
1724 Sacramento Street, Suite 200
San Francisco, CA 94109
Telephone: 415-552-0141
(See also Chapter 4 for a description
of this organization.)

The Society of Women Engineers
345 East 47th Street
Room 305
New York, NY 10017
Telephone: 212-705-7855

7

Space
and the World

Although some tend to think of space exploration as an arena in which the United States enjoys a unique position, this is an idea that will appear less and less true through the rest of the century.

In reality, it was never true—least of all during the early years of space development. The first spacecraft of all was the Russian Sputnik 1. Soviet spacecraft were the first to fly past the moon (Luna 1), the first to hit the moon (Luna 2), the first to soft-land on the moon (Luna 3), the first to land on Venus (Venera 3), the first to return photographs of Venus (Venera 9 and Venera 10), the first to land on Mars (Mars 2), and the first to return data from Mars (Mars 3). We tend to forget these facts, perhaps because of the phenomenal success of the U.S. spacecraft when they were finally directed to their planetary targets. The endless stream of pictures and data from Ranger, Surveyor, Lunar Orbi-

ter, Mariner, Pioneer, Viking, and Voyager have been un-
paralleled in the history of planetary sciences, and their
open dissemination provided striking proof of the sophistication of the U.S. unmanned programs.

The history of manned spaceflight and of space applications is similar. We are overwhelmed by the Apollo missions, by Skylab, and by the space shuttle. To many people
it comes as a surprise to learn that the first man to orbit the
earth was not Alan Shepard or John Glenn, but a Russian,
Yuri Gagarin. Or to find that the first woman in space, Valentina Tereshkova, was also Russian. Or perhaps to find
that the pioneering papers on communication satellites
were provided by an Englishman, Arthur C. Clarke, and
the first domestic communications-satellite system was developed by the Canadians.

Space has always been of interest to the whole world,
and today more and more countries are actively engaged in
space-program development, including joint international
space ventures. Every venture presents a wide range of career positions, so be sure to look at each technology, how it
can be applied, and where.

But what about job opportunities? They do exist in other
countries, and for people excited by the idea of living and
working in different countries, this idea should be seriously
considered.

There are a few things to note before rushing off to apply
for positions outside your national man-made boundaries.
First, academic training varies greatly from one country to
another, as do standards of living, customs, and the willingness to employ non-natives. We think this is likely to
change, as advanced technology continues to become more
and more international.

Second, if you do seek a job in another part of the world,
your chances are vastly better if you have scientific training.
If there is a universal language in use in the world, it is not
English, or Russian, or Japanese. It is mathematics, closely
followed by physics, chemistry, and engineering.

Third, all the rules that we emphasized in Chapter 5 apply even more to international jobs. Why should an employer from one country take a chance on someone from another country, whose academic credentials are difficult to verify, who probably has a language problem, and who may not be able to provide personal references of any value?

There are reasons, and you should make the most of them. Let's look at the reasons from an American viewpoint, noting that no matter where you are in the world, you have advantages you can identify. (Contact us if you need help.)

First, the technology of the United States continues to enjoy an excellent reputation, and in the field of space that reputation is probably higher than in any other field. Few countries express a desire to emulate the U.S. political system, police system, or legal system. But almost every country admires and would like to copy NASA and the U.S. space program. This reputation (whether you deserve it or not!) will work to your advantage, if you are an American citizen abroad. More and more countries, also, are developing fine space technology and application reputations.

Second, twenty years ago American salaries and living standards were the highest in the world. You could not hope to work abroad without taking a cut in pay. Today, many countries of Western Europe have higher per capita incomes than the United States.

Third, you may be able to turn what appears to be an obvious disadvantage abroad—English as your native (or second) tongue—to your own uses. English is today the principal language for scientific publication, worldwide. Even in Europe, many of the journals contain a high proportion of papers written in English. The ability to skim through such papers and offer rapid and accurate summaries of content is a real sales point to Japanese or German employers (but be aware, you will often find that their English is far superior to your German or Japanese; the average American's lack of foreign languages is perhaps even better known abroad than the technology).

There is one other thing to be said in favor of working in another country. It is great fun. If you enjoy travel, trying for a job in a different location on the planet is well worthwhile. And don't forget that space is only eighty miles from wherever you are.

The Canadian Space Program

We begin close to home, where the language problem is minimal and the lifestyle not too dissimilar.

Canadian efforts in space may seem small compared with those of Canada's southern neighbor. In fact, Canadian space programs are substantial, with expenditures of over one hundred million dollars a year planned for the period 1981–1985. This amount appears more impressive when we remember that the Canadians made a decision some time ago that they would buy launch services rather than develop their own. Almost all the budget is thus for science and applications. According to Canadian government statistics, over thirteen thousand men and women are employed in the Canadian space program, mostly by about sixty companies who provide space-support services. The principal government contractor is Spar Aerospace, in Toronto, which builds most of the Canadian satellites. Canadian estimates place the world market for Canadian satellite systems and services at over a billion dollars in the next ten years.

The Canadian space program is generally directed by the Ministry of State for Science and Technology. As we shall see, different parts are under the control of other specific agencies.

Canadian satellites have been flying for a long time. In September 1962, Canada became the third nation to develop and orbit a satellite (after Russia and the United States). This was Alouette I, for use in ionospheric research, and it had an unusually long lifetime. It continued to return data to Canadian receiving stations for more than ten years.

Given the country's great size (second largest in the world), it is natural that much of the space effort is focused on two problems: communications, and the measurement and management of natural resources.

REMOTE SENSING AND RESOURCE MANAGEMENT

This space activity is the general responsibility of Canada's Department of Energy, Mines and Resources. In 1972, at the time of the launch of the first U.S. Landsat spacecraft (at the time, known as ERTS-1) the department established the Canadian Center for Remote Sensing (CCRS) to manage national efforts, develop the necessary technology, and process resources data. These data are returned by the American Landsat, Seasat (which failed early and operated in 1978 only), and meteorological satellites, and will be returned eventually by the French SPOT satellites, the Japanese MOS and LOS satellites, the European Space Agency's ERS-1 satellite, and possibly others in the late 1980's.

CCRS operates image data receiving stations at Prince Albert, Saskatchewan, and at Shoe Cove, Newfoundland, which together provide complete coverage of Canada and incidentally of most of the United States. Received data are applied to Canadian problems of crop inventory, forest and wildlife management, sea-ice surveying, land and ocean mapping, and mineral and petroleum exploration.

To this point, Canadian activity has centered on the use of data provided by U.S. resources satellites. However, Canada has unique needs for monitoring conditions in the Arctic, and for this purpose a Canadian satellite, provisionally known as Radarsat, is under design for a late 1980's launch. It will carry a cloud-piercing synthetic aperture radar (SAR), to provide an all-weather coverage of Arctic and offshore ice patterns. The satellite will also be used to study land resources and offshore pollution.

The Canadian company of Macdonald Dettwiler & Associates has a worldwide reputation for the construction of satellite ground-receiving stations for earth-resources data. It is headquartered in Richmond, British Columbia, a suburb of Vancouver, but does work around the world.

For more information concerning the Canadian space remote sensing programs, contact:

> The Canadian Centre for Remote Sensing
> Department of Energy, Mines and Resources
> 2464 Sheffield Road
> Ottawa, Canada K1A 0Y7

COMMUNICATIONS

This was the earliest of the Canadian satellite programs, already an active area by the middle 1960's. Canada was one of the original eleven members of Intelsat (the International Telecommunications Satellite Organization), and in 1969 Telesat Canada was formed. This is a corporation that is operated jointly by the Canadian government and the private telecommunications companies, and it was created to develop and then provide service from a Canadian domestic satellite system. This began regular operation in 1973, with the launch of the ANIK A satellite. Telesat Canada was the first system of its kind in the world.

The whole ANIK series of communications satellites is controlled by Telesat Canada. The most recent, ANIK D, was launched by an American Delta vehicle in 1982, and more are scheduled to be launched by the shuttle. These satellites carry commercial messages and also advanced communications experiments. Although there are commercial communications satellites in use in the United States, there is nothing in this country quite like Telesat Canada— Comsat is the closest thing to it.

More advanced systems are now under development, including a Mobile Satellite Program (MSAT) that is intended to provide better communications with remote areas, particularly in the Canadian North. If you would like to combine a space career with a chance to see something of the Canadian Northern Territory, this would be the place to start! The MSAT system will provide terminals in places like the Arctic Islands, up about 75° north, where the Sverdrup Basin represents one of the world's largest, least explored, and most promising sedimentary basins—a possible location for great oil and gas reserves.

For more information on the Canadian communications satellite programs, write to:

> Department of Communications
> 300 Slater Street
> Ottawa, Canada K1A 0C8

SPACE SCIENCES

Since 1976, Canada has been working with the United States on equipment and experiments for use on the space shuttle. The most famous of these is certainly the Canadarm, the remote handling device that permits operations outside the space shuttle to be controlled by the astronauts inside. Other ongoing shuttle-related programs are instrument-package development and life-sciences experiments that will be carried on Spacelab; and over the longer term there are plans for a magnetometer package and possibly for Starlab, a one-meter aperture telescope for stellar exploration.

The Canadian space sciences programs are handled through a variety of establishments: the National Aeronautics Establishment has overall responsibility for the Canadarm development; the Herzberg Institute of Astrophysics in Ottawa handles atmospheric and interplanetary research programs; and the Canada Center for Space Science operates the Churchill Research Range in Manitoba and the Mobile Scientific Balloon Launch Facility near Winnipeg. In addition to NASA joint programs, there are also cooperative efforts with European countries, such as instrument development for the Swedish Viking satellite.

General information on the Canadian space sciences program can be obtained from:

> Canadian National Research Council
> 270 Albert Street
> Ottawa, Canada K1A 1A1

The European Space Agency (ESA)

At first sight, European space activities may be a little confusing. The European Space Agency has eleven member countries: Belgium, Denmark, West Germany, France, Ireland, the Netherlands, Spain, Sweden, Switzerland, and the United Kingdom. The member countries are required to support certain basic ESA activities, including science research and the overall ESA organization. In addition, members may take part in specific projects, with contracts being issued to companies in each country according to the proportion contributed to the project costs. Finally, any member nation may continue its own space program activities, separate from ESA, and several members do so on a substantial scale. These single-country activities will be described separately.

The total ESA budget is substantial. For 1982, it was almost three-quarters of a billion dollars. ESA programs span the field of space sciences, space applications, and space transportation systems, and offer today's prime competition with the space shuttle for commercial launch services.

MAJOR PROJECTS

The two activities of ESA that have taken the lion's share of available funds are Spacelab and Ariane.

Spacelab was built to provide a self-contained science laboratory that could be carried to and from orbit by the space shuttle. It can be thought of as a single, integrated, and very large module, which is fixed in the shuttle's cargo hold but which is not integrated into the orbiter's instrument and support systems. It contains both unpressurized and pressurized sections, and can thus be used with both shirt-sleeves and vacuum environments for conducting space experiments.

Spacelab is a sophisticated device whose development cost nearly nine hundred million dollars. NASA agreed when the development began that it would not pursue any competitive system. Since Spacelab forms a logical starting point for a space station, and since such a permanent station is now the first item on NASA's wish-list for the late 1980's, it is not surprising to find that ESA and NASA are both studying ways to employ Spacelab technology in permanent space-station design. During the Carter administration, when the U.S. space program was at an ebb, NASA was told to forget space stations and other advanced programs. However, in that period ESA continued to study the possible use of a free-flying Spacelab module as a space station.

Another project that grew out of Spacelab is EURECA, the European Retrievable Carrier. This is a package (unmanned) that can be placed in space by the shuttle, left there for up to six months of operation, and then retrieved.

At the same time as ESA was building Spacelab for use with the shuttle, it was also developing its own independent launch capability, in the form of Ariane. This is an "expendable" launch vehicle, like all the U.S. launches before the shuttle; thus it contains no reusable elements (the space shuttle reuses the orbiter and the solid-fueled strap-on rocket boosters, but not the external tank, which must be provided new for each launch).

One might think that it is much less expensive to have a system that can be used many times. Imagine the economics of an airline, for example, if the aircraft were thrown away after carrying one load of passengers or cargo—that's what you face with an expendable launch vehicle. In practice, however, the shuttle is still such a technologically complex and innovative system that it has no competitive cost advantage over the expendable vehicles. Thus, Ariane has a chance of capturing a good share of the market for commercial launches. Its U.S. competition is with Titan, Atlas, and Delta launch vehicles. Like them, it is liquid-fueled, with liquid oxygen and kerosene first stages, and a liquid hydrogen/liquid-oxygen third stage. This will give it the capability to place communication satellites weighing up to two and a half tons into high (geosynchronous) orbit, about the same as the shuttle can manage with its added expendable upper stage (the inertial upper stage, usually abbreviated to IUS).

To date, Ariane has been viewed as a vehicle for unmanned launches. Now, however, ESA is looking at the possibility of a "mini-shuttle," Hermes, which could be launched by an up-rated Ariane in the late eighties or early nineties. Hermes is a fascinating concept. It is small (forty feet long, with a wingspan of only twenty-four feet), and would weigh a total of only eleven tons. It could carry two crew and a thirty-two-hundred-pound payload into orbit, and then glide back to Earth landing after reentry in much the same way as the shuttle. For comparison, the much larger shuttle was designed for a maximum sixty-five-thousand-pound payload, but has not yet achieved that goal.

Hermes would just about fit into the shuttle's payload bay. On the other hand, Hermes could operate with orbital inclinations up to 60°, whereas the Shuttle Orbiter is designed for less than 50° inclination until launches from the Vandenberg Western Test Range are possible, probably in 1986. With the more inclined orbits, shuttle's maximum payload then drops to only thirty-two thousand pounds. ESA has also been considering an unmanned shuttle, Solaris, which would be a cargo-hauler for the space program, carrying raw materials to a processing facility in orbit and the finished products back to Earth.

Ariane is being handled as very much a commercial, worldwide launch service, through a government-industry consortium, largely French, known as Arianespace.

SPACE SCIENCE ACTIVITIES

A broad array of basic science missions is under development by ESA. These include:

1) The International Solar Polar Mission. This is a spacecraft that is designed to fly over the sun's poles and explore the particles and fields there. It became something of a *cause célèbre* for U.S.–European relations when the U.S. government, having first agreed to participate in the program with one of two satellites, then backed out of its commitment because funds were not available. The Europeans will go ahead with a single-satellite mission, scheduled for 1986 launch. But the experience left a bitter taste in European mouths, and will certainly hinder future collaborative programs.

2) EXOSAT. This is an X-ray astronomy satellite, carrying a pair of X-ray telescopes and other X-ray detectors. It is designed to occupy a very eccentric orbit, so that the moon can be used to occult X-ray sources and so allow very accurate locations to be determined for them.

3) Giotto. This will be a probe to investigate Halley's Comet. It is scheduled for 1985 launch, and will carry a color camera and eight other instruments for analysis and observation of the comet's tail and nucleus.

4) Hipparcos. This will be a satellite designed to map the positions and motions of one hundred thousand stars with very high accuracy. It is scheduled for a launch late in 1986.

5) Five projects are now being studied for possible late eighties or early nineties launches. These are:

• AGORA—the asteroid gravity, optical, and radar analysis mission.

- FIRST—the far infrared, submillimeter-wavelength telescope.
- SOHO—the solar high-resolution observatory.
- XMM—the X-ray multi-mirror observatory.
- CLUSTER—an array of four formation-flying satellites to measure the magnetosphere.

6) A variety of other projects are also under consideration, including a Mars orbiter, a solar interior-studies satellite, a Magellan ultraviolet observatory, and a timing/transit satellite.

The wide variety of these missions points to the vigor and vitality of the European science programs. One reason for ESA's program stability is probably the method by which funding is obtained. Since a number of countries must commit to providing part of the budget, any one of them is constrained against backing out once a first commitment has been made. Thus, there is less chance of the random year-to-year chopping and changing of budgets that characterizes the progress of U.S. programs as they see-saw between Congress and the administration. We will return to this problem in the next chapter, and look at its consequences in more detail.

SPACE APPLICATIONS ACTIVITIES

1) Materials processing in space is an important part of Spacelab's planned programs. Spacelab 1 will carry materials science experiments, and in 1985 the Spacelab D-1 mission purchased by West Germany will be devoted to low-gravity experiments and to materials processing. This will be a next evolutionary step following the long series of low-gee experiments known as the TEXUS series of sounding rocket flights.

2) ESA has launched a continuing series of communications satellites, beginning in the 1970's with the OTS-1

test satellite. This continued with the ECS-1 and ECS-2 operational European Communications Satellites, with 1982 and 1983 launches; the MARECS maritime communications satellite (MARECS-B was lost in a failed launch of the Ariane rocket); and the Large Communications Satellite (L-SAT) designed to provide direct broadcast television to France and West Germany and to carry other communications experiments.

3) The Earthnet system operated by ESA was set up to receive earth resources data from the U.S. Landsat series of spacecraft, and will after 1984 receive data from the French SPOT earth-resources satellite. International weather data are currently provided by the Meteosat 1 and 2 satellites, which from geostationary orbit give coverage of almost a whole hemisphere of the earth. An ESA ocean monitoring satellite, ERS-1, is also being developed, to carry a radar altimeter, radar scatterometer, and synthetic aperture radar in low-earth orbit.

As this brief summary should have suggested, the ESA programs are numerous and complex, and include lots of career positions. If you wish to obtain additional materials concerning any one of them, one approach is to write to ESA headquarters in Paris. However, we have a different suggestion. ESA operates a Washington, D.C., office, with knowledgeable and helpful staff, and we recommend that inquiries from within the United States be directed there. The address is:

> European Space Agency
> 955 L'Enfant Plaza
> Suite 1404
> Washington, D.C. 20024
> (Telephone: 202-488-4158)

The Japanese Program

The Japanese space program has a fairly complicated structure. The Institute of Space and Astronautical Science (ISAS) is the oldest space group of Japan. It is a multi-university institute, under the direction of the Ministry of Education, and it was responsible for orbiting the country's first satellite, OHSUMI, in 1970. The man behind ISAS is the recognized leading figure of Japanese space development, Hideo Itokawa. Now in his seventies, and indisputably one of the great figures of the world's space program, Itokawa founded the Kagoshima Space Center in southern Kyushu Island. It was his work and leadership that led Japan in 1970 to become the fourth nation to orbit its own satellite on the LAMBDA 4-S launch vehicle. This took place just two months before a launch by mainland China.

However, ISAS is not primarily responsible for launch-vehicle development or for applications satellites, though it does have its own launch facility on the Osumi Peninsula in southern Kyushu. Most launch development and all applications satellites come under NASDA, the National Space Development Agency of Japan. NASDA, the "Japanese NASA," was established in 1969, an outgrowth of the National Space Development Center. Finally, NASDA, ISAS, and all other Japanese space activities are coordinated by a space activities commission, a five-member group that reports to the Japanese prime minister.

Perhaps the easiest way to think of ISAS and NASDA is to regard ISAS as responsible for space sciences, NASDA as responsible for space applications, and both as having a launch capability.

In practice, this is just the beginning of the complications. To understand the Japanese space program, it is also necessary to understand the unusual relationship between government and industry in that nation. In the United States, government and industry often regard each other almost as adversaries, with opposing objectives. Certainly we have no difficulty here distinguishing the roles of the two groups. In an oversimplified picture, it is government that provides monitoring and control of industrial actions, and industry that is responsible for manufacturing and entrepreneurial activities.

Things are ordered differently in Japan. The three giants of the Japanese space business are Nippon Electric Company, Mitsubishi Electric Corporation, and Toshiba Corporation. They work very closely with the Japanese government, and it is common to find industrial staff members assigned for periods to work as part of government agencies. This interlocking of people and programs will become apparent when we describe the development of launch vehicles. There also seems to be far more intellectual respect in Japan for government workers, probably because the government employs the cream of the crop from the Japanese university system.

To add the final layer of complication, each of the three main Japanese space companies has formed close affiliations with different American groups. Nippon works closely with Hughes Aircraft in Los Angeles and with RCA Astro Electronics in New Jersey, Mitsubishi works with Ford Aerospace, and Toshiba works with General Electric at Valley Forge, Pennsylvania. Many of the parts for Japanese satellites are built in this country, and sometimes whole satellites. The Himawari-1 meteorological satellite, for example, was built by Hughes in California, and launched on a Delta rocket from Cape Canaveral.

Against this confusing background, we will now try to describe the current activities of each component of the Japanese space program.

NASDA

The administrative headquarters are in Tokyo's World Trade Center building, but the technical centers are elsewhere. The main research and development facility is the Tsukuba Space Center. It is on the main island of Honshu and about fifty miles northeast of Tokyo. The launch facility is the Tanega-shima Space Center in southwest Japan, located on Tanega-shima, a small island south of the main western island of Kyushu and about 120 miles southeast of Nagasaki. There is also a rocket-engine development facility at Kakuda on the northeast shore of Honshu, and a Landsat ground-receiving station near Tokyo.

NASDA works in four main areas:

1) Launch-vehicle development. The main support contractors are Mitsubishi Heavy Industries and Nissan Motors, and the vehicles are built in Nagoya, 160 miles west of Tokyo and Japan's third largest city. Two series of launch vehicles exist, the N series (which closely resembles the American Thor-Delta vehicle), and the newer H series, which will carry satellites weighing half a ton and more to geosynchronous orbit. A new LE-5 liquid-hydrogen/liquid-oxygen second stage is being built and tested at the Kakuda facility.

2) Communications satellites. Japan has orbited a series of these, beginning with ones launched for them by NASA from Cape Canaveral. However, the latest series (CS-2a and CS-2b: Communications Satellites 2a and 2b; and BS-2a and BS-2b: Operational Broadcasting Satellites 2a and 2b) will be launched on N-series rockets from Tanega-shima, in the period 1983–86. The direct-broadcast satellites will beam television programs to the half-million Japanese households beyond the range of conventional television transmitters.

3) Applications and test satellites. These include the weather satellites (Himawari-2 and Himawari-3, both designed for geostationary orbit); the Marine Observation Satellite, MOS-1, which will be launched in 1986 or 1987 and will be used among other things to assist the Japanese fishing fleet; the Earth Resources Satellite ERS-1, which is expected to carry a synthetic aperture radar and other imaging systems, to be used mainly for

oil and mineral exploration and assessment, and probably be launched in 1989; and test payloads that set the stage for larger projects in areas such as space materials processing.
4) Manned space research. It seems probable that Japan will perform space materials processing on board the space shuttle in 1987 or 1988. These would be performed by a Japanese payload specialist, giving the country's first manned presence in space.

In addition to this, Japan is making plans for its own manned program. NASDA in 1982 released the design of a small shuttle, which is in many ways similar to the European Space Agency's Hermes proposed spacecraft. The Japanese vehicle is 46 feet long, has a 24.5-foot wingspan, weighs only 20,000 pounds, and would carry a crew of 3, plus half a ton of payload. Rather than gliding back in to a landing like the shuttle, the Japanese craft would fly back on turbo-jet engines. There was no discussion as to the boost phase that would carry the Japanese shuttle most of the way into space.

Farther out in time and much more ambitious is an unofficial report of a Japanese design for a manned communications platform in geosynchronous orbit. To be economical, such a station calls for development of a reusable space tug that can carry passengers and payloads between low-earth orbit and geosynchronous orbit. If a manned geosynchronous platform is truly in the Japanese plans for the future, they are thinking of space development on the largest scale.

ISAS

The early work of the Institute of Space and Astronautical Science goes back to 1955, and thus predates the launch of the first artificial satellite. The organization began projects in scientific rocketry under Hideo Itokawa's direction immediately upon its formation. Since 1970, ISAS has averaged a launch per year of science satellites, many under the direction of professors from Tokyo University.

Some of the major projects include:

1) The Hakucho (Swan) satellite, launched in 1979. This orbiting astrophysical observatory provides radio, optical, and X-ray observations, and has been operated in conjunction with M.I.T. in this country.
2) The Honitori (Firebird) satellite, launched in 1981, and designed to study the phenomena of solar flares.
3) The Planet-A Halley's Comet probe.
4) A principal-investigator role for particle acclerator experiments on Spacelab.
5) Sounding rocket programs, using launch vehicles designed by ISAS and built at Nissan Motors. Although ISAS has in the past clung to its own launch program, in the late eighties it seems almost certain that all Japanese launches will be consolidated in NASDA's activities. However, ISAS will undoubtedly be an important factor in their design.
6) Deep space programs, including construction of a sixty-four-meter antenna that will be used to track and control the Halley's Comet missions and subsequent planetary program spacecraft.

The Japanese program is being conducted in the same quietly aggressive and perfectionist spirit that has marked their entry into the steel, automobile, and electronics industries. For someone seeking a space career with added elements of the exotic, Japan is an excellent place to look. However, one of the easier ways to do so might be through the American companies mentioned earlier, who are engaged in joint activities with Nippon, Mitsubishi, and Toshiba. In addition to these programs, McDonnell Douglas has a team working in Nagoya with NASDA contractors.

For more information on the Japanese space programs, we recommend that you do not write directly to Japan. Instead, send your request for information to the Japanese Embassy in this country, at the following address:

NASDA, Embassy of Japan
The Watergate, Suite 900
600 New Hampshire Avenue, N.W.
Washington, D.C. 20037
(Telephone: 202-234-2266, Extension 210)

The Swedish Program

This is a country where the language problem is minimal. English is spoken by so many people in Sweden, and spoken so well, that a visiting American has little difficulty in working with Swedish professionals. Sweden's standard of living is high, and average salaries exceed those in the United States. The country is now placing great emphasis on the growth of high-technology industries, with space one of the most active fields. Most developments are funded by government research programs or international collaborative programs.

SPACE COMMUNICATIONS

With the country's population of only eight million spread over a large area, communications is naturally one of the first fields of interest for space projects. Sweden, together with Norway and Finland, now plans to launch the experimental communications satellite Tele-X as a precursor to a possible Nordsat direct broadcast satellite (Nordsat has been discussed for some years by Sweden, Norway, Denmark, and Finland, but intergovernmental negotiations have so far blocked its progress).

Tele-X will be launched by an Ariane rocket in 1986. It will serve multiple purposes, as follows:

• Transfer of digital data, with direct computer-to-computer links

- High quality image transmission, for remote newspaper printing, transmission of aerial and space images, and map production
- Service as telephone switching systems
- Transfer between picture telephones
- Mobile telex links, between trucks, coaches, trains, and aircraft
- Electronic mail

Sweden's participation in Tele-X will be led by Svenska Rymdbolaget, the Swedish Space Corporation (SSC), headquartered in Tritonvägen, in the Stockholm suburb of Solna. SSC is a state-owned corporation, part of the Ministry of Industry, and heavily involved in most Swedish space activities. Sweden's great interest in space communications also derives in part from the presence there of the international telecommunications giant L. M. Ericsson, which already supplies telephone networks around the world and is headquartered in Sweden. However, there will also be substantial financial support to Tele-X from France and Germany, and the satellite will be built around a basic structure developed for use in French and German spacecraft.

THE VIKING SATELLITE
This will be Sweden's first satellite, built by Saab-Scania under SSC management, and with Boeing as subcontractor for the satellite platform. It is planned for 1984 launch, and like Tele-X it will ascend on an Ariane vehicle (the same one that will be carrying the French SPOT satellite). Viking weighs about half a ton, and is designed for research into particles and fields close to the earth. It will particularly be used to study the particles that give rise to the aurora borealis, and the magnetic fields that control their movement. In addition to instruments to measure the particle and field densities, Viking will carry an ultraviolet camera that can take space pictures of the aurora.

The Viking spacecraft will be exploring the radiation belts, regions of intense fields around the earth, and its possible rapid degradation in this energetic environment is a cause for concern. The satellite's planned lifetime is only about six months, though experimenters have prepared for and are hoping for a year of data.

THE KIRUNA FACILITY

All data from Viking will be returned to the ESRANGE receiving station and processing facility at Kiruna, in the extreme north of Sweden and, at 68° north, well into the Arctic Circle. This northerly position is a definite advantage, since the station is able to receive data from polar-orbiting satellites on each of their revolutions.

Kiruna is a center for space research as well as a tracking and processing operation. The station already receives Landsat data, and will be one of the principal receiving stations for SPOT data. A new image-processing center for earth-resources data is now under development there, in addition to the command center that will provide control of the Viking spacecraft.

Kiruna is also the headquarters for EISCAT, an international collaboration and a major research center for European studies of the aurora. EISCAT (European Incoherent Scatter Scientific Association) operates large radar systems that provide a detailed analysis of the plasma densities found in the earth's ionosphere and magnetosphere. The financial support for the program comes from Sweden, Norway, Finland, Britain, France and Germany.

More information concerning the Swedish space program can be obtained from:

> Svenska Rymdbolaget
> Tritonvägen 27
> S-171 54 Solna
> Sweden

Alternatively, inquiries from within this country may be directed to the Swedish Embassy, at this address:

> Science Attaché
> Swedish Embassy
> 600 New Hampshire Avenue, N.W.
> Washington, D.C. 20037
> (Telephone: 202-298-3500)

The French Program

The French government sees space as an important part of its overall plan to strengthen the French economy and international position, through an emphasis on high-technology activities. The French have therefore emphasized both national and international space projects, and this gives their program a rather complicated three-part structure. There are ongoing activities as part of the European Space Agency, there are bilateral cooperative efforts with the United States, Russia, West Germany, and Sweden, and finally there are national programs that are wholly French.

The French space agency is C.N.E.S. (Centre National d'Etudes Spatiales). Legally established in 1961, it began operations in 1962 and has its headquarters in Paris and major facilities in Toulouse, Evry, and Kourou (Guiana). About half of the C.N.E.S. budget goes to ESA programs, 20 percent to bilateral efforts, and the remaining 30 percent for the national program.

The ESA programs have already been described, but we should add that the French are the prime contractor for the Ariane Program, and provide for the launch of the Ariane rocket at their Guiana facility. Arianespace, a commercial organization set up in 1980 to sell the services of Ariane, is owned almost 60 percent by France (Germany is the second largest stockholder, with nearly 20 percent).

BILATERAL COOPERATION

C.N.E.S. is working with NASA in the following areas:

• Spacelab, where French experiments in material sciences, the study of growth processes in space, and extragalactic research will be performed. The French will also modify their Echograph equipment, already used for medical monitoring on the Salyut spacecraft, for operation on Spacelab.
• Participation in the Galileo project, with two French experiments.
• Participation in Space Telescope work.
• High-resolution imaging of the sun, using equipment to be flown on U.S. sounding rockets.
• Geodesy experiments, requiring the installation and French operation of four geodetic stations (at Kerguelen, Tahiti, Kiruna, and Ougadougou).
• Experiments on the Upper Atmosphere Research Satellite (UAR), to study winds and temperatures between 85 and 250 kilometer altitudes.

French-Russian space cooperation goes back to 1966, when an intergovernmental agreement was signed for the study of space for peaceful purposes. The Soviet Union and France have the following cooperative programs:

• Participation in the Russian manned program, via the guest cosmonaut program (one French "spationaut," Jean-Loup Chrétien spent a week on the Salyut 7 space station in mid-1982).
• Study of the magnetosphere using the ARCAD III sensor package. The Soviet Oreol 3 satellite carrying this was launched in September, 1981, and the experiment is expected to provide data until 1987.
• Experiments on the Soviet VEGA mission, which involves a Venus lander and a Halley's comet fly-by. Launch is scheduled for 1984.

• A variety of scientific experiments, including gamma-radiation studies (satellite launch in 1984), primate-behavior observation in space, ultraviolet stellar-spectrum measurement, and studies of the solar wind and the magnetosphere.

France and West Germany plan the launch of direct broadcast satellites in 1985. One geosynchronous satellite will cover French needs, one German needs, and there will be provision for a back-up spacecraft. Project management will be performed by an integrated team of French and German engineers based in Munich, and a development contract with the Eurosatellite consortium permits Belgian and Swedish industrial cooperation in the project.

France also participates in the Swedish Viking satellite development.

FRENCH NATIONAL PROGRAMS

Although France has ongoing scientific and communications satellite programs, the most ambitious of the French projects is the development of the SPOT (Système Probatoire d'Observation de la Terre) earth-resources satellite, scheduled for 1984 launch by Ariane. SPOT carries a high-resolution observing system, providing ten-meter resolution black and white images, and twenty-meter color and infrared images. (The best resolution of comparable U.S. spacecraft is the thirty-meter system of the Landsat-4 Thematic Mapper).

SPOT will provide pictures of most of the earth's surface on a regular basis. It will use a multiple linear-array sensor, not yet flown on any U.S. civilian satellite, and the mirrors that direct images to the linear array can be turned away from the vertical. The satellite can thus be used to construct stereoscopic images from overflight data taken on successive days, it can take a second look at an area within 3 to 5 days (compared with 16 or 18 days for Landsat), and it can look 475 kilometers to either side of the vertical (compared with 92 kilometers for Landsat).

France regards the SPOT system as a commercial, fully operating system, in contrast to the United States earth-resources satellite programs, which have always functioned in a research mode. SPOT 2 has already been approved for production, and there are longer-range plans for two more satellites in the series. To aid sales of SPOT data around the world, C.N.E.S. in mid-1982 formed SPOT IMAGE, headquartered in Toulouse. SPOT IMAGE will have offices in the United States and elsewhere, and will sell earth-resources images and tapes on a commercially competitive basis. In addition to recording of data onboard the SPOT spacecraft for subsequent transmission to Kiruna (Sweden) and Toulouse, arrangements are being made to receive data directly at a set of ground receiving stations around the world. Applications of the data and training of users will be performed through a government-commercial consortium known as G.D.T.A. (Groupement pour le Développement de la Télédétection Aérospatiales), located in Toulouse.

Although predominantly French, there are also international cooperative elements of this program. Sweden will provide the SPOT onboard computer, and Belgium the power supply subsystem.

Additional information concerning the French space activities may be obtained from these locations:

> Scientific Attaché for Space Affairs
> Embassy of France
> 2011 Eye Street, N.W.
> Washington, D.C. 20006
> (Telephone: 202-659-3550)

> C.N.E.S.
> 129 Rue de l'Université
> 75007 Paris
> France

The British Program

Discussion of British efforts in space is complicated by the fact that no single agency has main responsibility for the country's space activity. There is also no defined British policy toward the role of space in national affairs or its relationship to the general development of British technology.

And yet, paradoxically, Britain has been continuously and actively involved in space projects since the early 1960's, beginning with the Ariel series of scientific satellites. Britain pioneered the development of the first European operational communications satellite for military use, Skynet II; it took a leading part in the creation of the European Space Agency in 1975; and the country now spends about 80 million pounds (nearly $150 million) a year on space, and provides more financial support to European Space Agency activities than its relative gross national product would require.

One reason for the complicated situation is the number of different parties involved. Somewhat independent and loosely linked space activities are pursued by the Ministry of Defence, the National Environmental Research Council (NERC), the Science and Engineering Research Council (SERC), the Ministry of Information Technology, and the Royal Aircraft Establishment at Farnborough. In addition, various British companies such as British Aerospace and Marconi Space and Defence Systems are significant contractors to the European Space Agency.

The British situation may change rapidly in the next year or two. The idea of an integrated space program has been widely discussed since early 1982, and a big step forward was taken in May 1983 with the announcement of a coordinated national remote-sensing program and the establishment of a National Remote Sensing Programme Board, which will pull together all the agencies mentioned. There is strong support in Britain today for a space program that is focused more on purely national projects and less committed to ESA activities (at present, 75 percent of the British space budget goes to ESA programs).

The British space program includes a substantial role in ESA activities, notably the experimental OTS communications satellite, the European Communications Satellite (ECS), the Maritime Communications Satellite (Marecs), the Large Communications Satellite (L-Sat), the Ariane launcher (Britain has a much smaller share here than France and West Germany), the ERS-1 earth-resources satellite, and the Meteosat 1 and 2 weather satellites.

In addition, Britain will have its own direct-broadcasting satellite system with service projected to begin in 1986, a military communications satellite, Skynet IV (to be operational by 1985 as the successor to Skynet II), and planned ground stations for data reception from the ERS-1, Landsat, and SPOT earth-resources satellites. Projected space expenditures over the next few years show a doubling of today's 80-million-pound annual budget.

For additional information on the British space program, write to:

> Department of Industry
> Space Branch
> 123, Victoria Street
> London, SW1E 6RB
> England

or in this country to:

> Dr. Anthony Cox
> Counselor for Science and Technology
> British Consulate
> 3100 Massachusetts Avenue, N.W.
> Washington, DC 20008
> (Telephone: 202-462-1340)

The Indian Program

Despite formidable economic problems, India has elected to develop a strong domestic space program. In fact, India looks to space technology to help in solving some of the country's biggest obstacles to progress: a polyglot citizenry and an undeveloped technological infrastructure.

LAUNCH VEHICLES

The Indian program began early, in 1963, with a small launch facility at Thumba in southern India. Seven satellites have so far been launched from this site, and over the next decade a billion-dollar program will develop, launch, and use a series of earth-resources and communications satellites. This will be done using an Indian-made four-stage rocket, the SLV-3, and also a series of larger launch vehicles that are scheduled for development in the middle and late 1980's.

It was the SLV-3 that launched India's first experimental Rohini satellite, in July 1980, from the eastern missile range at Sriharikota, and it brought the country into that select group of nations with the capability to achieve orbit using their domestic satellites. The SLV-3 is a solid propellant rocket, similar in size and design to the U.S. Scout rocket. The next launch vehicle will be much larger in scale, permitting the launch of payloads up to 330 pounds. This is scheduled for 1984 completion. The next step calls for a booster that can place one-ton applications satellites into polar orbit.

APPLICATIONS SATELLITES

Although an Indian cosmonaut will participate in a Russian-Indian mission in the near future, India has made it clear that neither manned spaceflight nor planetary exploration is any large part of their program. The Indian Space Research Organization (ISRO) has stated the national goal: "to discover the country from space." Applications satellites are the highest priority, with education, resources management and monitoring, and country-wide communications heading the list. In the words of ISRO's chairman, Professor Satish Dhawsan, "The main aim of the Indian space program is towards the development of communications and Earth observation satellite systems suited to India's needs."

Following the launch of an experimental satellite, Aryabhata, in 1975, pursuit of these objectives began in earnest four years later, with the development and orbiting of the Bhaskara-1 earth-observation satellite. This carried a TV camera and a microwave sensor, for data collection in the disciplines of hydrology, forestry, mineral geology, and oceanography. An improved version of the same satellite was launched in 1981.

The next generation will be the Remote Sensing Satellite, known as IRS, which is scheduled for launch aboard a Russian booster in 1985. IRS will carry multispectral scanners, similar to those used on the U.S. Landsat satellites, and also a multiple linear array, similar to that to be carried on the French SPOT satellite. It is intended for primary use in agriculture, forestry, meteorology, and hydrology. India already operates a ground receiving station for receipt of Landsat data, at Hyderabad in central India, and the same facility will be able to receive IRS images and scientific data.

In the field of communications satellites, India placed the APPLE satellite in orbit in July 1981, aboard the European Space Agency's Ariane rocket. This was a natural next step after the initial experiments in satellite instructional television, carried out with the U.S. ATS-6 satellite and with the German/French Symphonie satellite. Both these projects were coordinated through the western receiving station in Ahmedabad.

The Indian space program suffered a setback in 1982, when the INSAT-1A satellite was abandoned after a series of problems. This village communications satellite was to have combined the three functions of telecommunications, direct broadcast television, and weather watching. The successor, INSAT-1B, will be launched using the space shuttle within the year.

For more information on the Indian space program, we suggest that the best point of contact from within the United States is directly to the embassy. The address is:

> Educational Attaché
> Embassy of India
> 2107 Massachusetts Avenue, N.W.
> Washington, D.C. 20008
> (Telephone: 202-265-5050)

The Chinese Program

We should say at the beginning that we do not have immediate expectations that you'll find a job in China's space program, but anything you want you should go for. Chinese-American cooperative efforts in space do exist (for example, a U.S. company will construct and install the Chinese Landsat receiving station), but these are relatively small. However, we expect China to be an increasing influence in space affairs in the rest of this century. We include a brief description of their progress and plans, and those of the USSR, to round out the world picture of space development.

China's first satellite was launched in April 1970. It thus became the fifth nation, after the USSR, the United States, France, and Japan, to orbit an artificial satellite using a launch vehicle developed domestically. Since then it has conducted a dozen successful launches, using its FB-1 booster. According to Chinese reports, four of the payloads were recovered, in a region of southwest China bordering the Tarim Basin. All missions to date have been unmanned. Although the Chinese disclose little information about their space technology and less about their mission objectives, these unmanned satellites are suspected of being for purposes of photo-reconnaissance. The Chinese space program has a civilian component, but it serves predominantly military purposes. The recovered payloads were thus likely to be film canisters of military targets or areas of high strategic interest.

The ability to return and recover payloads suggests that China is well on the way to a manned program. This is supported by films shown to a U.S. shuttle team. The crew of STS-3, Gordon Fullerton and Jack Lousma, were shown movies of suborbital tests on dogs and rats. Other Chinese sources have referred to astronaut training now being conducted on selected Chinese pilots. Before the end of the 1980's, a Chinese manned presence in space seems likely.

China is also developing more sophisticated launch systems. One reported booster is said to use liquid-hydrogen upper stages, which would make it a very advanced instrument. It will be used to place weather and communications satellites into geosynchronous orbits.

The Soviet Program

If there is little information available about the Chinese program, the Soviet program deserves a book to itself. The USSR has been continuously active in space since the first launch, that of Sputnik 1 in October 1957. Space is a top priority for the USSR, and the Soviet people have tremendous interest in all space projects, seeing a large potential for the solution of problems here on Earth. Soviet astronauts (cosmonauts) are still regarded as great national heroes, in a way that in the United States was reserved for the original seven of the Mercury Program and the first men on the moon.

Like the Chinese program (or, for that matter, the U.S. program) the Soviet space effort is highly influenced by military and political goals. This has two main effects. First, there has been a steadily increasing real expenditure on space, variously estimated as from less than 1 percent to as much as 8 percent per year. This should be compared with the major fluctuations in U.S. space spending discussed in Chapter 2.

Second, the Soviet program does not have an "open skies" component, comparable to NASA in the U.S. program (although much of the U.S. space program is also classified). Thus, many elements of Soviet space activity are not announced to the rest of the world and must be deduced from other evidence. For example, the architect of the Soviet space program until 1966 was Sergey Pavlovich Korolev, a master organizer and engineering genius whose contributions should be ranked with those of Tsiolkovski, Goddard, Oberth, and Von Braun. Prior to his death, his name was never mentioned in any official statement on the Soviet space program.

The statements given in this section are thus drawn partly from official Soviet announcements, but a good deal also comes from the work of close students of the Soviet space program, such as Craig Covault, Nicholas Johnson, James Oberg, Saunders Kramer, and the late Charles Sheldon, who have painstakingly pieced together a fairly complete picture from fragments of information. (Recommended reading on this subject are *Red Star In Orbit,* by James E. Oberg, Random House, 1981, the *Handbook of Soviet Manned Space Flight* and *Handbook of Soviet Lunar and Planetary Exploration,* by Nicholas L. Johnson, American Astronautical Society Science and Technology Series, Volumes 47 and 48, 1979 and 1980, and *Soviet Space Programs 1966–70, 1971–75, and 1976–80* [3 separate volumes] by Charles S. Sheldon III, Marcia S. Smith, *et al.*, Congressional Research Service of the Library of Congress. See Chapter 2 for full references.)

A few of the long list of significant Soviet "space firsts" were mentioned at the beginning of this chapter. Another index to recent Soviet activity in space is provided by the fact—known beyond doubt—that the Soviet Union launches more satellites each year than the rest of the world *combined.* Each year, three hundred to four hundred tons of payload are launched by the Soviets, compared to little more than a tenth of that amount by the United States. Even allowing for the generally greater sophistication of U.S. satellites, and hence a need for fewer spacecraft, the Soviet launch activity is most impressive.

Today's most significant space activities of the Soviet program, ongoing and planned, appear to us as follows:

1) Since the launch in 1977 of the Salyut-6 spacecraft, the Soviets have had what amounts to a permanent orbiting space station. Salyut-6, and its successor, Salyut-7, have logged over forty-five thousand man-hours of cosmonaut occupancy, including several missions of over six months duration by Soviets, and shorter "guest cosmonaut" visits by Czechoslovakian, Polish, East German, Bulgarian, Hungarian, Vietnamese, Cuban, Mongolian, Rumanian, and French astronauts. With dual docking ports, the Salyut space stations can be supplied in orbit, even when unoccupied, by an unmanned Progress spacecraft. The next manned crew (flying the Soyuz series of spacecraft) can dock at the other port and then unload the two and a half tons of supplies provided by Progress. There is every evidence that long-term space occupancy is a permanent goal of the Soviet program, and that medical worries that originally arose over the physical effects of long periods of weightlessness have proved unfounded.

2) The Salyut stations appear to be the precursors of more ambitious stations, to be launched in the late 1980's. These larger stations are reported to be planned for permanent occupancy by about twelve cosmonauts. Although no details have been released, an exciting statement of intention was made in 1979, when chief cosmonaut Shatalov said: "We are close to the constant operation of orbital stations—to around-the-clock,

year-round work of cosmonauts aboard them, with replacement of crews . . . and regular delivery of necessary materials."

3) To make possible the efficient construction of these larger space stations, a more powerful "superbooster" is believed to be under development. Such a "G-class" (this is a U.S. designation, not a Soviet one) launch vehicle will be able to orbit payloads of over 100 tons (some estimates place the number as high as 200 tons), comparable with the payload capacity of 125 tons of the largest U.S. booster ever employed, the Saturn 5.

Such a booster does more than make possible the large low-earth orbit stations. It would permit, as one example, the placing of a Salyut-class spacecraft in orbit around the moon. Soviet statements concerning lunar-orbiting stations have been made in recent years, although this is usually discussed in terms of assembly in Earth orbit of boosters and Salyut-style components, without discussing the G-class booster. It is unwise to attach a date to the large booster, since its imminent appearance has been prophesied for fifteen years.

Soviet plans do not appear to stop there. Tsiolkovski's writings have served to set the long-range goals of the Soviet space effort, and he dreamed of exploring the moon and the planets with manned spacecraft. For over a decade, Soviet cosmonauts and scientists have echoed that dream, to the point where Valery Ryumin, the Soviet cosmonaut who has already spent almost a year in space, declared himself quite ready to go: "If an expedition to Mars were being prepared and it should be necessary to hold a year-long stay in space as an intermediate step, I think that we would readily agree to such work." Soviet scientists speak confidently of such a mission taking place before the year 2000. The experience gained so far with the Salyut station suggests that, without major changes, it could already serve as a suitable module for a round trip to Mars, perhaps ten years from now.

4) Closer to the present, the Soviets are said to be working on a reusable winged spacecraft similar to the United States shuttle orbiter. It has been under development for at least five years, with projected launch date within the next few years. The first flights are expected to take place as the payload for an expendable launcher, with only the winged spacecraft initially recovered. However, in the 1990's or earlier, the Soviets are likely to move to a fully reusable system, replacing the expendable first stage with a flyback booster capable of reuse along with the spacecraft.

A reusable vehicle is badly needed to keep down the costs for supplying the Salyut space stations, and for launch and return of cosmonauts. Economic analysis suggests that the availability of a reusable craft for Salyut 6 supply would have saved nearly five billion dollars in that project alone.

5) On scientific missions, the stated 1980's priorities for the Soviet program are as follows:

- The study of the earth's upper atmosphere, magnetosphere, and Sun-Earth relationships.
- Planetary studies with unmanned vehicles, including analysis of planetary atmospheres and chemistries.
- Continued spaceborne experiments in materials processing and in development of in-space industries.
- Global monitoring systems for natural resources and atmospheric measurement.
- Cometary studies, particularly missions to Halley's Comet, which will be close to the sun in 1986. The Soviet program calls for a fly-by at a distance of a few thousand kilometers, photographing the comet's nucleus and analyzing dust and gases found in it.
- A large space telescope, with an aperture stated as "sev-

eral meters," to be launched within the next ten years.
- More sophisticated weather and communications satellites, including development of multipurpose communications satellites.
- Research on new methods of propulsion for travel in space (as compared with travel to and from space). These include solar sails, ion drives, and nuclear and plasma engines.

6) On the military side, a ground-based antisatellite (ASAT) weapons system that started its development in the late 1960's in reaction to the U.S. ASAT weapons has been tested with only about 50 percent test successes. Recently, improved unmanned reconnaissance satellites for both land and ocean surveillance have been developed. These include all-weather vehicles equipped with spaceborne radar, plus film-recovery systems that image targets for about a month, then eject film canisters for reentry. Since the present Soviet systems appear to have a total imaging lifetime of only a few weeks each, the new developments will greatly decrease the launch requirements (about thirty Soviet imaging reconnaissance spacecraft are now launched per year). The Soviet official reports say nothing at all about these developments (the United States is equally closemouthed on the subject of its own classified reconnaissance and space weapons systems); thus our comments on Soviet military space activities have come wholly from U.S. sources. In August 1983, Soviet leader Yuri Andropov committed to halting ASAT tests in a unilateral moratorium as long as no other country tests, to dismantling existing ASATs, and to returning to the table to negotiate a treaty to ban all space weapons.

The overall picture of the Soviet space program is very impressive. There is a consistency of funding and, even with inevitable technical problems like those of developing a very large booster, a steady progress toward space exploration and utilization. It is a pity that cooperative efforts of the space superpowers are restricted to "guest cosmonauts" in Soviet programs, and such activities as Spacelab in the United States program. If we were to seek to define a first-rate near-term cooperative effort for the United States and Soviet space programs, it could be done in a single sentence. *The United States has a reusable space transportation system (the space shuttle) but nowhere to go, and the USSR has a space station* (Salyut) *and no reusable transportation system to supply it with people and equipment.* And we can develop a joint satellite-security-monitoring system based upon existing systems.

Is it conceivable that we might, some time this decade, see these two needs matched, and a cooperative program defined? Today, this is still over the horizon. Although we have some cooperation in unmanned programs (for example, in developing and using COSPAS and SARSAT, search-and-rescue satellite systems that allow rapid location of travelers in trouble), we have no approved plan for any coordinated manned flight activity, similar to the 1975 Apollo-Soyuz Test Program where a cosmonaut and an astronaut shook hands in space.

However, it is this exciting type of international cooperation that we should keep firmly in mind when we look at space around the world. A global cooperative program for manned exploration of the moon, Mars, and the universe has tremendous popular appeal, with few apparent military-weaponry implications. As cooperative space ventures develop, more jobs will be created. In the words of the USSR's deputy director of the Institute for Space Research, Yuri Zaitsev: "The detailed study of even the closest neighbors of Earth like Venus and Mars, let alone the more remote planets of the solar system, requires considerable funds. It would be expedient, therefore, to develop further international cooperation in space research. . . . The success of the joint Apollo-Soyuz flight in 1975 shows that given mutual understanding it is possible to implement major scientific projects, thus promoting international detente and consolidating peace. This flight was of great historic value, for it symbolized an improvement in Soviet-American relations on the basis of peaceful coexistence . . ."

A combined effort involving the U.S. space shuttle and the Soviet space stations; and, cooperative monitoring of the earth, or joint exploration of the planets. Wishful thinking? Perhaps—but the sort of thinking that we prefer to the "Us versus Them" attitude still playing so large a part in space developments around the world. It is clear that space belongs to all world citizens, and that we shall work in space together.

8

Into the Future

"For I dipped into the Future, far as human eye could see,
"Saw the Vision of the world, and all the wonder that
    would be."
                                    —Alfred, Lord Tennyson
                                    "Locksley Hall," published 1842

"And then, the earth being small, mankind will migrate
into space, and will cross the airless Saharas which separate
planet from planet and sun from sun. The earth will become
a Holy Land which will be visited by pilgrims from all the
quarters of the Universe. Finally, men will master the
forces of Nature; they will become themselves architects of
systems, manufacturers of worlds."
                                    —Winwood Reade
                                    The Martyrdom of Man, published 1872

"The future cannot be predicted, but futures can be invented."
                                    —Dennis Gabor, 1963

The Long Road

In this book we have been concentrating on practical matters: How do you train for a job involving space, where are the jobs to be found, how do you apply for them, how do you get yourself *involved* in space? Now it is time to take a look a little further ahead. Jobs are important, work is important, career paths are important—but to make a real contribution to your life, you have to know where those jobs finally lead, not just this year and next year, but for the rest of your long life and if possible on beyond that, to the next centuries.

It's not easy to take the long perspective. Everyday living occupies the biggest part of our attention. To most of us, a year is a long time, a big chunk of our lives. Worrying about tomorrow, we rarely think of what will be happening twenty years from now. Still less do we try to imagine the world as it will be in a few centuries. We cannot even imagine what it might be like in a thousand years, though the changes to human existence will probably be bigger than those in the previous five thousand years.

The effort to take the long view is more than just worthwhile; it is absolutely essential, for it provides answers to a crucial question: How should we best use our time and energies, *now*? As Dennis Gabor said, the future cannot be predicted, it can only be *invented*. Since we know humans are moving into space whether we like it or not, we want to participate in that invention, rather than being swept along into tomorrow as complacent bystanders.

The present battle of ideologies between East and West is a good example of the need for a long-term view. That struggle has been going on for most of this century, since long before the end of the Second World War. It would be easy to take the short view, and imagine that it could go on forever. And from the month-to-month viewpoint that most of us follow, most of the time, it certainly feels that way. We are weary of cold wars, Us-and-Them rhetoric, remorseless arms build-up, and endless tension. Many are working to ensure that earth's arms race does not spread into space.

But if we think in terms of centuries, rather than months or years, all human experience tells us that the present political situation will be no more than a fleeting moment of history. The mightiest and most stable empires the world has ever known, that which grew up along the fertile Nile Valley six thousand years ago, or the great power of Rome, extending its military might over Europe for a millennium, are gone. Every one faltered, fell, and became at last a relic of history.

It is hard to believe that our present preoccupations will be any more lasting. One day the United States itself along with the thirteen colonies will be a piece of history, with the issues that concern us so much today as unfamiliar to our descendants as the wars of the Spanish succession or the complicated interactions of the ancient Greek city states.

Most of the things we do today will not appear even as footnotes in future histories, but there is reason to believe that one activity the human race has undertaken in the past quarter of a century will continue to be of importance as long as civilization continues. Humankind will explore and develop space, and will remember the generations when that exploration began.

This is a self-indulgent thought, and one that should normally be dismissed as arrogant. We have just pointed out how most events of the past diminish in time, so that we do not remember the names of the pharaohs, or recall the sequence of Roman emperors or Chinese dynasties. Since the time of Copernicus, we have gradually been forced to accept the idea that our planet occupies no unique position in the universe. We have seen first the earth, then the sun, and finally the galaxy, displaced from the central place in the cosmos. The lesson should be applicable to time as well as space. We should be reluctant to conclude that the period of our own lives is in any way significant in a larger sense. All the same, we cannot resist that assertion. We are living at a uniquely important moment of history. What has happened in the past generation, and what happens in the next, will be recognized a thousand years from now as a critical point in the history of humanity.

The crisis may be of opportunity grasped or of opportunity lost. We are the victims of a curious coincidence of timing. At the same time as we have developed the capability to move off this planet and begin the exploration of space, we have also gained the power to destroy civilization. We have enough weapons in our stockpiles to make Earth uninhabitable to the human species. In 1920, long before nuclear war offered its special potential for self-destruction, H. G. Wells said that human history was more and more a race between education and catastrophe. Today, it appears as a race between international understanding and extinction.

Arthur Clarke, writing in 1950, before the space age and in the first generation of nuclear energy, stated the alternatives:

"There is no way back into the past; the choice, as Wells once said, is the Universe—or nothing. Though man and civilizations may yearn for rest, for the Elysian dreams of the Lotus Eaters, that is a desire which merges imperceptibly into death. The challenge of the great spaces between the worlds is a stupendous one; but if we fail to meet it, the story of our race will be drawing to its close. Humanity will have turned its back upon the still untrodden heights and will be descending again the long slope that stretches, across a thousand million years of time, down to the shores of the primaeval sea."

Since then the choice has become clearer, and new space-weapon potential has added a new urgency. The descent that Clarke envisioned may not take place as a long steady decline, through the quiet twilight of the human race. Our fall could be completed, like that of Milton's

Satan, in the space of one day. There would be no twilight, only the noon glare of hydrogen fusion.

In a warring world, the open frontier of space takes on a stronger significance. The past ten years have provided a new way to reach that frontier, more sophisticated space-borne tools to verify arms control treaties, and a better understanding of the space environment; but in that same period we have seen other developments that should cause alarm. We believe that the United States space program is in trouble. Let's see why we think so.

Problems for the U.S. Space Program

"I know that some knowledgeable people fear that although we might be willing to spend a couple of billion dollars in 1958, because we still remember the humiliation of Sputnik last October, next year we will be so preoccupied by color television, or new-style cars, or the beginning of another national election, that we will be unwilling to pay another year's installment on our space conquest bill. For that to happen—well, I'd just as soon we didn't start."
—Hugh L. Dryden

"Where there is no vision, the people perish."
—Proverbs, 29:18

In the past couple of years, we have seen some sensational results from the U.S. space program. The space shuttle, after what seemed like endless delays and setbacks, finally made it into space and proved itself a unique space transportation system, one that probably impresses everyone else in the world far more than it impresses people in the United States. We in the United States have become spoiled by success, to the point where we take it for granted.

And while the shuttle was opening up a new era for travel to and from low-earth orbit, the Voyager spacecraft were probing the farther reaches of the solar system, sending back incredible pictures and scientific data of Jupiter, Saturn, and their moons. In the past twenty years, our space vehicles have flown to, and sent back images from, Mercury, Venus, Mars, Jupiter, and Saturn. These are all the planets of the solar system that were known at the time of the American Revolution. Uranus, Neptune, and Pluto were discovered in the eighteenth, nineteenth, and twentieth centuries respectively. Just two hundred years passed from the discovery of Uranus by William Herschel in 1781 to the fly-bys of Jupiter and Saturn by Voyager. The past twenty years have justifiably been termed the golden age of planetary exploration, a time that comes only once to any science.

Yet despite all these successes, we claim that the American space program is in trouble. The nature of that trouble becomes obvious as soon as we look at the 1980's, and compare them with the 1960's and 1970's. On both the manned and the unmanned fronts, the United States space effort has been in retreat for some years. The successes we have described began their development in the late 1960's and early 1970's. It took more than ten years to develop the space shuttle, and an equal length of time to move the Voyager project from conception to planetary observations. If we want to know what similar programs we can expect in the eighties and early nineties, lead times are so long on major projects that we have to look around us now for their beginnings.

What do we see? Since 1976, there have been almost no new projects approved for the U.S. civilian space program. It is easy to list many that have been proposed and rejected:

1) Halley's Comet comes close to the sun only once every seventy-six years. That will happen in 1986, and presents a unique opportunity to take a close look at some of the original material from which the solar system was formed, probably in close to its original condition.

A United States program was proposed several years ago, and not approved. It looks as though there will be Japanese, Soviet, and European missions to explore the comet, but no American mission.

2) Skylab was America's closest approach to a space station. Astronauts occupied it for periods of up to eighty-four days, in three missions between May 1973 and February 1974. The project was an engineering and scientific triumph, including an unexpected and very successful in-space repair and installation of a sunshade.

A permanent manned station was the logical next step. However, the United States flew no manned missions of any kind from July 1975, when the joint U.S.–U.S.S.R. Apollo-Soyuz Test Project was conducted, until the first launch of the shuttle in April 1981. During that same period, the Soviet space program's Salyut spacecraft were used for long periods of continuous occupation by cosmonauts. One Russian, Valery Ryumin, has spent almost a full year in space, in two six-month missions on board the Salyut 6 spacecraft.

Even with the shuttle in full operation, extended missions are not proposed. Today's standard missions are only planned to last 7 days, though studies and proposals have been made for payload packages that could extend missions to 30 days, or even to 180 days. The idea of the permanently manned space station remains NASA's most important goal, but that new national space objective has yet to be accepted by Congress and the administration. Meanwhile, there has been for some time talk of a twelve-man permanent Soviet station planned for 1985 launch.

3) The planet Venus is permanently shrouded in clouds. The radar carried by the Pioneer Venus-probe spacecraft permitted a look through those clouds, but the detail provided in the images was not good. A closer look was proposed in the Venus Orbiting Imaging

Radar (VOIR) mission, which would have given maps of Venus with detail as small as one kilometer. After all the preliminary design work had been completed, the project in 1981 failed to find the necessary budget.

4) In 1972, the United States launched a satellite known as ERTS—the Earth Resources Technology Satellite. This was designed to measure and monitor phenomena on the surface of the earth, and it was the first space-craft ever launched explicitly for use in earth-resources observation. In 1975, it changed its name to Landsat-1, and since then three more spacecraft have been placed in orbit, Landsat-2 in 1975, Landsat-3 in 1978, and the more advanced Landsat-4 in 1982.

The scanners and vidicons of the Landsat satellites provide images of every part of the earth between lati-tudes 81 North and 81 South. For more than ten years, the telemetry systems on these spacecraft have sent back information of great value in agriculture, geo-logical exploration, forestry, pollution monitoring, water resources, rangeland management, desert stud-ies, and general land use. Today, however, the U.S. program appears ready to die. After the launch of one more spacecraft, Landsat-D', there is nothing more in the pipeline. The administration killed the funding for two further satellites, and currently does not even have the launch funds for Landsat-D'. Ironically, this comes at a time when other nations are expressing great inter-est in earth-resources satellites. The French will launch their SPOT satellite in 1985, the Japanese plan a launch later this decade, and the European Space Agency will launch their ERS-1 spacecraft, probably in 1985. By the end of the 1980's, the United States will be bystanders in a program whose potential they were the first to recognize and demonstrate to the rest of the world.

5) The world has been using up its fossil fuels at a great

rate for the past century and a half. Coal, oil, and gas that were tens of millions of years in formation are being consumed so fast that shortages appear inevitable in the next century—unless we can find a renewable energy source. Solar energy is one of the most promising possibilities. In 1968, Peter Glaser proposed a space-based approach to solar-power generation. Large orbiting collectors (kilometers across) would convert solar energy to microwave radiation and beam it down to receivers on the ground. There it would be transformed to electrical power and used in electrical power grids.

The possible advantages of such an approach are many—twenty-four-hour a day power production, from a nonpolluting and nondepletable energy source—but the technical problems needed to build solar-power satellites are formidable. We need to launch (or mine in space) large quantities of construction materials. We also need to develop a full-scale space construction industry. The solar-power satellite is thus not a short-term solution to the world's energy problems; but it could be a superior long-term answer, available in thirty or forty years' time when oil, gas, and coal supplies begin to dwindle. For this to be possible, the project should now be in its design stages—we are talking of a multibillion-dollar program, ten times the size of the Apollo Program, with correspondingly long lead times and a final large national commitment (and since we are interested in jobs, let us remember that Apollo at its peak provided employment for over four hundred thousand people).

A few years ago, the first U.S. government studies of the concept were conducted by the Department of Energy. These analyses did not show the idea to be impossible or impractical. In fact, the studies discovered no problems or difficulties that could not be addressed with *today's* technology, still less with that we should have available early in the next century. It was logical to proceed to the next step, of more complete satellite system design and costing.

Instead, all government funding has ended. There is no longer a solar-power satellite program in this country, other than that conducted by private groups, such as the Sunsat Energy Council, who are so convinced of the project's importance that they continue despite government indifference.

6) Rockets that use chemical fuels have exhaust velocities of only a few kilometers a second. Since it is the exhaust velocity that determines the ratio of payload to fuel mass (the higher the exhaust velocity, the more payload that can be carried for a given amount of fuel), other propulsion systems are preferable to chemical rockets, particularly for low-thrust missions that carry unmanned satellites around the solar system.

Electric propulsion is one of these preferred systems. It offers exhaust velocities of up to 70 kilometers a second, compared with the 4.5 kilometers a second limit of a liquid hydrogen/liquid-oxygen rocket. The Solar Electric Propulsion System (SEPS) was proposed for development in the 1980's. It would be of great value in many missions for planetary exploration. However, all funding for SEPS was canceled in 1981. There are now no projects for advanced nonchemical propulsion systems in the United States space program.

It would be misleading to suggest that there will be no new starts in the next decade. The Venus Radar Mapper, planned for a 1980's launch, will be a smaller, lower-cost version of the canceled VOIR. Galileo, a Jupiter orbiter and probe, is still proceeding, though at a slower pace than originally planned. An orbital X-ray Astrophysics Facility (AXAF) is moving steadily forward, as is the Advanced Communications Technology Satellite, a NASA/industry combined effort. And the space telescope, scheduled for 1986 launch, should change forever our ideas about the way to perform astronomical observations (we may never again be willing to settle for a ground-based view through the earth's atmosphere).

In the next few years, we will surely see more new project starts. But even so, on balance we are looking in the 1980's at a period of retreat from the ambitious and vigorous programs of the sixties and seventies.

If we want the situation to change, we must first understand its causes. Then we will look at the mechanisms that exist to create changes, and see how an individual can exert maximum effect on those mechanisms.

Linkages: Space and the United States

There is a tendency to speak of "the space program" as though it were some separate and isolated entity, thriving or dwindling without affecting or being affected by other developments in society. That is a dangerous and an unrealistic idea. The development of the space program, and the ability to carry through successful space projects, is intimately linked with the overall state of the economy. It is perhaps even more closely linked with the general state of science and technology programs.

It seems unlikely that we can ever look forward to a thriving space program unless the economy is doing well. And today's economic progress is increasingly the product of and responsive to technological progress. To close the logical loop, we cannot therefore look for a healthy space program unless the other advanced technologies are doing well. Diminish overall technological progress, and you harm the space program. Stimulate that progress, and you help the space program. The relationship is as simple as that.

This should be no surprise to us. Space is the high frontier, certainly; but it is (and has always been) the *high-tech-*

nology frontier, or it is nothing. You don't get to space by rubbing two sticks together, you get there applying the best devices that modern science can devise and modern technology can fabricate.

For this reason, a strong space program is impossible without strong support from many areas of technological development—computers, aeronautics, communications, electronics, materials, metallurgy, medicine, engineering, and scores of others. To say that one is in favor of a strong space program but opposed to overall technology development is a form of self-delusion.

This point is of central importance because we have seen in the United States in the past few years a retreat from technology. It has become the suspected source for all our woes, the natural place for us to point a finger of blame when things go wrong. We are surrounded by problems that seem to be the direct result of our technology, from radioactive waste to polluted lakes, acid rain, and toxic landfills. There is emphasis on "appropriate" technology, which is often interpreted to mean the smallest possible amount of it. As that attitude has become more popular, the overall economic health of the United States has suffered a decline. We believe that this is not coincidence; attitude to technology and economic progress are directly linked. Minimal technology is certainly possible—a hundred nations around the world prove it every day, against their will. Unfortunately, that minimal technology is not something on which we can ever build a space program. When a country's research and development efforts decrease, the space program is obliged to shrink with them.

Perhaps we can take a lesson from history. More than twenty years ago, soon after this country committed itself to the Apollo Program, the magazine *Newsweek* listed a score of issues such as hunger, disease, and pollution that they gave more priority to than the "Buck Rogers stunt" of landing a man on the moon. Two decades later, the same publication noted that the post-Apollo decline in U.S.

space funding has coincided with a drop in the nation's technological productivity, and in its economic strength. That drop has been matched by an increasing emotional and psychological malaise throughout the nation.

Coincidence? Perhaps. But perhaps that look toward open frontiers is an absolute essential. Ben Bova, in his book *The High Road,* quotes a suggestive historical analogy:

"Nearly 500 years ago the Chinese developed deep-water navigation to the point where their ships ranged from Madagascar to the East Indies, exploring, trading, tapping the wealth of the Indian Ocean and west Pacific. But abruptly the Ming Dynasty halted this flourishing commerce. No one really knows why, except that the central Chinese government decided to keep its people tightly bound to their own shores. Within a century China was being picked apart by Europeans, who found a backward, ignorant nation that was already fragmenting into petty principalities."

A hundred years from now, will twenty-first-century traders from Japan or Europe sell their technical wizardry to the Divided States of America, where technology slowed and stagnated in response to a loss of national will?

The Military Program

Loss of faith in technology is one danger facing the space program. Second, and no less important, we must examine the possible implications of our military activities. It is no secret that the United States space program today has a bigger military component than ever before. We don't object to the military in space—only to an expanded space weapons program.

The military space program is certainly nothing new. NASA and the military programs make use of the same set of launch vehicles, and have excellent transfer of information between them. Although we read a great deal about NASA, and little or nothing about the U.S. military space activities, military and intelligence satellites have been flying since the first years of the space program. They serve a uniquely important function: stabilizing of the situation among the world's superpowers. Satellite systems help to reduce the level of uncertainty that nations have regarding their own and each other's activities. With high-resolution imaging satellites, large industrial developments, armament build-ups, and troop movements are impossible to disguise. These imaging spacecraft are the principal tool for disarmament or arms control treaty verification, and are regarded by many people as the biggest justification for the space program thus far. President Johnson stated that this single reconnaissance use had paid for the whole space program, many times over. At the same time, satellite communications help to keep commanders in minute-by-minute and second-by-second contact with bases and fleets around the world, minimizing the chance of accidental military engagements or unauthorized unilateral actions. Thanks to the military space program, security systems are in place to achieve peace right now.

To look for a historical analogy, by the late 1960's we had come to the stage in space military development that was reached by tethered balloons 150 years ago for ground-combat situations. Manned balloons could be used for reconnaissance and to report on enemy actions, but they were not able to be used as weapons.

Today, we face the next stage, historically similar to the development of aircraft and dirigibles. These could be used to attack directly enemy territory and possessions. We are not quite in that position yet. Although the Soviet Union and the United States are both working on the development of methods of destroying satellites in low-earth orbit, space is not yet an arena for battle, nor are there powerful weapons in space. However, it takes little imagination to envision the same sort of eventual proliferation of space weaponry, country by country, as we have seen in nuclear weapons development.

That danger was recognized early, and back in the 1960's the United States and Russia signed a treaty banning weapons of mass destruction from space. This treaty has generally been interpreted to apply to nuclear weapons, but not to conventional weapons—or to very unconventional ones, such as particle beams and lasers. Antisatellite weapons, or ASATs, non-nuclear devices (the United States dismantled the nuclear-tipped warheads on the first U.S. ASATs due to the electromagnetic radiation pulse research that proved U.S. satellites would also be destroyed as well as other satellites if a nuclear blast occurred in space) designed to destroy a single satellite, do not clearly violate the treaty. When we realize this "single satellite" that may be destroyed could include a ballistic missile, we see the clear basis for a new arms race, this one to be conducted in space. The objective would be a spaceborne system, "hardened" against attack from other antisatellite weapons, that could destroy a missile launched from anywhere on earth. This has led people to conjecture that the next step after an antisatellite system would be an anti-antisatellite system, then an anti-anti-antisatellite system, and so on, countermeasure after countermeasure, in a never-ending sequence.

To add a little to the complexity, any satellite we can now launch to low earth orbit (below about five hundred miles) is very vulnerable to *ground-based attack*. A simple sounding rocket, with a payload of gravel, would destroy an orbiting satellite. The relative velocity of the two, meeting at several miles a second, would give each pebble the force of a high-speed bullet. Some security would be offered by moving the satellite to the "high ground," up in geosynchronous orbit (twenty-two thousand miles up, where a satellite's orbital period is twenty-four hours) or beyond. However, this raises the level of technology needed for a protected ASAT, and adds another escalation clause to an unwanted orbital arms race.

The value of a military space program to the security and peace of mind in the United States and other countries appears to us to be beyond question. Any development that stabilizes the relationships of the world and diminishes the chance of war is valuable almost beyond measure. Without a vigorous military, even the superpowers would not last for a year.

At the same time, because the military potential of space now stands at a major new threshold—of direct action rather than observation—there is a corresponding major new danger. In our urge for military parity and access to "the high ground," it would be absolutely disastrous to our long-term interests if the new frontier of space became exclusively or primarily an arena for combat. The military space program cannot be our only or our dominant effort. "Space" suggests emptiness, but beyond the earth the solar system contains huge resources, probably more than we can now guess at. These words are worth committing to memory:

"We all know that it has more than three times as many mountains, inaccessible and rocky hills, and sandy wastes, as are possessed by any State of the Union. But how much is there of useful land? How much that may be made to contribute to the support of man and of society? These ought to be the questions . . . the agricultural products of the whole surface . . . never will be equal to one half part of those of the State of Illinois; no, nor yet a fourth, or perhaps a tenth part."

This sounds like Senator Proxmire, or some other antispace representative, arguing against exploration of the moon or Mars. But it is actually Daniel Webster, addressing the United States Senate on June 27, 1850. He was talking not about the moon, but about California.

The solar system will be the California, the new frontier, for the twenty-first century. The potential riches of space, with boundless supplies of energy and raw materials, are humanity's last best hope for universal prosperity. If that window to the future were to be confined to military uses, especially to weapons, a century from now our descendants might stand on a depleted, polluted, exhausted earth, and curse the folly of their ancestors. We must find ways to keep that window open to peaceful exploration and exploitation, and to the creation of a new peace with an abundance of space-related careers.

In fact, we can work toward converting the war industry into a space industry. The same people who build missiles and bomber planes can build spacecraft and space habitats. Everyone keeps his or her job, and new jobs will be created. New markets and businesses will emerge.

How to Help the U.S. Space Program

We have discussed two great dangers that face the U.S. space program in the 1980's: retreat from technology and concentration on military needs. Both these emphases reflect deep-seated fears of the unknown. To counterbalance them, positive forces must be applied to help the civilian space program through difficult years in the 1980's. Although country-wide grass-roots space support is essential, the single most important place to apply force was identified early in this book: Washington, D.C. Let us look at the pressure points for action. You can educate and influence people at any step in the process.

Before a new NASA project (or a program of any other government agency) can begin, a long and complicated process must be completed, in which the executive and legislative branches of the U.S. government engage in a gigantic and cumbersome *pas de deux.*

It begins in the spring, when NASA's (let us focus attention on this one agency) top officials present to the President their suggestions and estimated costs for the year-after-next's programs. This will normally be a mixture of ongoing programs and proposed new starts, and typically it asks for a good deal more than the agency expects to receive.

The President, with substantial assistance from the Office

of Management and Budget (OMB), reviews NASA's budget request along with that of every other agency, and adjusts it to achieve the administration's preferred balance of programs in terms of their perceived balance of national priorities. NASA officials contribute to this first iteration, making their best case with the White House and with OMB. The resulting final budget is then submitted to Congress by the President at the end of January, to request funds for the following fiscal year (the government fiscal year runs from October 1 to September 30).

There may be little that you as an individual can do to influence the process to this point. However, submission of the "final" budget to Congress is not the end of the story—it is little more than the beginning.

In Congress, the budget is cut up into pieces and each section handed over to the appropriate committees of the House and the Senate, for their review. In this operation the Congressional Budget Office and the Senate Budget Committee play a role similar to that of OMB in the executive branch of the government.

First, the *Authorization Committee* analyzes the proposed budget of each agency. It is this group's job to authorize the programs that each agency may undertake, and often to add or delete items. Their recommendations are brought before the relevant full committee for review. Even when the full committee has examined and decided on the final form for programs, this does not give funding to an agency. The agency's proposed budget is also under review by the *Appropriations Committee,* which will prepare views and estimates for the full committee, recommending the spending of a certain level of funds.

The views and estimates of both Authorization and Appropriations Committees are presented to the Budget Committees by about the end of March, and during the January–March period the committees and their subcommittees hold hearings on the proposed budget. In April, there is an opportunity for a presidential update on the budget, and another such opportunity in July.

These activities are going on in both the House and the Senate, which must reconcile their separate proposed budget approvals; but that agreement will not be reached for a while yet. By the end of August, the authorizing bills and appropriation bills are passed, and money is at last appropriated for every agency's activities by approval of a bill by the full chambers of each house.

We have not reached the point yet where the agency can spend even one dime. The bill is now sent back to the President. He may sign it, but as often as not he will veto it. If he signs it, the process terminates and we have a budget. If he vetoes it, Congress has a choice. It can override the veto by a two-thirds majority vote of both houses, in which case the budget becomes law; or it can send the budget proposals back to the committees, in which case the whole authorization and appropriation review cycle will begin again. The Senate and House will modify their budget approvals as they think fit, and seek to reconcile between them. In practice, there is continuous negotiation between the administration and both houses of Congress, hoping to avoid the extreme measures of veto, override, or stalemate.

If this sounds like a complicated procedure, be assured that it is far more complex than our short description would suggest. In addition to the general flow of the budget from executive to legislative to executive, every congressman, congresswoman, and senator has a constituency to represent, and their own items they wish to see added to or struck from the budget. The result is a flow chart that no single sheet of paper could capture. During the year, there are also various opportunities to seek supplemental budgets for programs, or to exchange one item for another. All dates are approximate. The government usually begins the new fiscal year with no budget approved and signed, and is obliged to run for a while on a "continuing resolution," where expenditure rates cannot exceed those of the previous year.

Where can you be heard in all this activity? Well, the individual stands little chance of influencing the administration's attitude at the time of budget preparation, though there have been occasional successes with massive write-in campaigns. The best hope lies in influencing the positions of the House and Senate committees, and particularly of the individual congresspeople, senators, and staff aides. Every congressman and congresswoman, and every senator, is sensitive to opinions voiced by the electorate. These men and women are the pressure points that we have been looking for.

If you want to give a boost to the U.S. space program, Congress and the administration are the places to make yourself heard. (And don't forget to call industry presidents and chairpeople of the boards.) Tell them you prefer to see an exciting space program, with no weapons in it. The following lists give the composition of the Senate and House committees most relevant to space-program authorization and funding. Other committees (such as Foreign Affairs and Armed Services) also have considerable influence on space matters.

SENATE COMMITTEES

In the Senate, the most relevant committees are the Commerce, Science and Transportation Committee (which has responsibility for space), the Appropriations Committee, and to a lesser extent the Energy and Natural Resources Committee. Their composition is as follows:

APPROPRIATIONS

Democrats	Republicans
John C. Stennis (Mississippi)	Mark O. Hatfield* (Oregon)
Robert C. Byrd (West Virginia)	Ted Stevens (Alaska)
William Proxmire (Wisconsin)	Lowell P. Weicker (Connecticut)
Daniel K. Inouye (Hawaii)	
	*Chairman

Democrats	Republicans
Ernest F. Hollings (South Carolina)	James A. McClure (Idaho)
Thomas F. Eagleton (Missouri)	Paul Laxalt (Nevada)
Lawton Chiles (Florida)	Jake Garn (Utah)
J. Bennett Johnston (Louisiana)	Thad Cochran (Mississippi)
Walter D. Huddleston (Kentucky)	Mark Andrews (North Dakota)
Quentin N. Burdick (North Dakota)	James Adbnor (South Dakota)
Patrick J. Leahy (Vermont)	Bob Kasten (Wisconsin)
Jim Sasser (Tennessee)	Alfonse M. D'Amato (New York)
Dennis DeConcini (Arizona)	Mack Mattingly (Georgia)
Dale Bumpers (Arkansas)	Warren Rudman (New Hampshire)
	Arlen Specter (Pennsylvania)
	Pete V. Domenici (New Mexico)

COMMERCE, SCIENCE AND TRANSPORTATION

Democrats	Republicans
Ernest F. Hollings (South Carolina)	Bob Packwood* (Oregon)
Russell B. Long (Louisiana)	Barry Goldwater (Arizona)
Daniel K. Inouye (Hawaii)	John C. Danforth (Missouri)
Wendell H. Ford (Kentucky)	Nancy Kassebaum (Kansas)
Donald W. Riegle (Michigan)	Larry Pressler (South Dakota)
J. James Exon (Nebraska)	Slade Gorton (Washington)
Howell Heflin (Alabama)	Ted Stevens (Alaska)
Frank R. Lautenberg (New Jersey)	Bob Kasten (Alaska)
	Paul S. Trible (Virginia)
	*Chairman

(It is interesting to note that Senator John Glenn, third American into space and first American to orbit the earth, does not serve on this committee. He serves on the Foreign

Relations Committee, the Governmental Affairs Committee, and the Special Committee on Aging. He remains, however, an influential figure on Senate activities pertaining to space.)

ENERGY AND NATURAL RESOURCES

Democrats	*Republicans*
J. Bennett Johnston (Louisiana)	James A. McClure* (Idaho)
Henry M. Jackson (Washington)	Mark O. Hatfield (Oregon)
Dale Bumpers (Arkansas)	Lowell P. Weicker (Connecticut)
Wendell H. Ford (Kentucky)	Pete V. Domenici (New Mexico)
Howard M. Metzenbaum (Ohio)	Malcolm Wallop (Wyoming)
Spark M. Matsunaga (Hawaii)	John W. Warner (Virginia)
John Melcher (Montana)	Frank H. Murkowski (Alaska)
Paul E. Tsongas (Massachusetts)	Don Nickles (Oklahoma)
Bill Bradley (New Jersey)	Chic Hecht (Nevada)
	John H. Chafee (Rhode Island)
	John Heinz (Pennsylvania)

*Chairman

Space affairs are the particular province of the Subcommittee on Science, Technology and Space, which is itself part of the Committee on Commerce, Science and Technology. The subcommittee is chaired by Senator Slade Gorton, who took over in January 1983, following the defeat of the previous chairman, astronaut Harrison Schmitt, in his bid for reelection. Senators Barry Goldwater, Nancy Kassebaum, and Paul Trible have been staunch supporters of the U.S. space program.

HOUSE COMMITTEES

In the House of Representatives, the most relevant committees are the Appropriations Committee and the Science and Technology Committee. Of the latter, the Natural Resources, Agriculture Research and Environment Subcommittee, the Science, Research and Technology Subcommittee, and the Space Science and Applications subcommittee are particularly involved in space affairs. The composition of these committees is as follows:

APPROPRIATIONS

Democrats	*Republicans*
Jamie L. Whitten* (Mississippi)	Silvio O. Conte (Massachusetts)
Edward P. Bolan (Massachusetts)	Joseph M. McDade (Pennsylvania)
William H. Natcher (Kentucky)	Jack Edwards (Alabama)
Neal Smith (Iowa)	John T. Myers (Indiana)
Joseph P. Addabbo (New York)	J. Kenneth Robinson (Virginia)
Clarence D. Long (Maryland)	Clarence E. Miller (Ohio)
Sidney R. Yates (Illinois)	Lawrence Coughlin (Pennsylvania)
David R. Obey (Wisconsin)	C. W. Bill Young (Florida)
Edward R. Roybal (California)	Jack F. Kemp (New York)
Louis Stokes (Ohio)	Ralph Regula (Ohio)
Tom Bevill (Alabama)	George M. O'Brien (Illinois)
Bill Chappell, Jr. (Florida)	Virginia Smith (Nebraska)
Bill Alexander (Arkansas)	Eldon Rudd (Arizona)
John P. Murtha (Pennsylvania)	Carl D. Pursell (Michigan)
Bob Traxler (Michigan)	Mickey Edwards (Oklahoma)
Joseph D. Early (Massachusetts)	Bob Livingston (Louisiana)
	Bill Green (New York)
	Tom Loeffler (Texas)
	Jerry Lewis (California)

*Chairman

Democrats	*Republicans*
Charles Wilson (Texas)	John Edwards Porter (Illinois)
Lindy Boggs (Louisiana)	Harold Rogers (Kentucky)
Norman D. Dicks (Washington)	
Matthew F. McHugh (New York)	
William Lehman (Florida)	
Jack Hightower (Texas)	
Martin Olav Sabo (Minnesota)	
Julian C. Dixon (California)	
Vic Fazio (California)	
W. G. (Bill) Hefner (North Carolina)	
Les AuCoin (Oregon)	
Daniel K. Akaka (Hawaii)	
Wes Watkins (Oklahoma)	
William H. Gray III (Pennsylvania)	
Bernard J. Dwyer (New Jersey)	
William R. Ratchford (Connecticut)	
William Hill Boner (Tennessee)	
Steny H. Hoyer (Maryland)	
Bob Carr (Michigan)	
Robert J. Mrazek (New York)	

SCIENCE AND TECHNOLOGY

Democrats	Republicans
Don Fuqua* (Florida)	Larry Winn (Kansas)
Robert A. Roe (New Jersey)	Manuel Lujan, Jr. (New Mexico)
George E. Brown, Jr. (California)	Robert S. Walker (Pennsylvania)
James H. Scheuer (New York)	William Carney (New York)
Richard L. Ottinger (New York)	F. James Sensenbrenner (Washington)
Tom Harkins (Iowa)	Judd Gregg (New Hampshire)
Marilyn Lloyd Bouquard (Tennessee)	Raymond J. McGrath (New York)
Doug Walgren (Pennsylvania)	Joe Skeen (New Mexico)
Dan Glickman (Kansas)	Claudine Schneider (Rhode Island)
Albert Gore, Jr. (Tennessee)	Bill Lowery (California)
Robert A. Young (Missouri)	Rod Chandler (Washington)
Harold L. Volkmer (Missouri)	Herbert H. Bateman (Virginia)
Bill Nelson (Florida)	Sherwood L. Boehlert (New York)
Stan Lundine (New York)	Alfred A. McCandless (California)
Ralph M. Hall (Texas)	Tom Lewis (Florida)
Dave McCurdy (Oklahoma)	
Mervyn M. Dymally (California)	
Paul Simon (Illinois)	
Norman Y. Mineta (California)	
Richard J. Durbin (Illinois)	
Michael A. Andrews (Texas)	
Buddy MacKay (Florida)	
Tim Valentine (North Carolina)	
Harry M. Reid (Nevada)	

*Chairman

Democrats

Robert G. Torricelli (New
Jersey)

Frederick C. (Rick) Boucher
(Virginia)

SPACE SCIENCE AND APPLICATIONS SUBCOMMITTEE

Democrats	*Republicans*
Harold L. Volkmer* (Missouri)	Manuel Lujan, Jr. (New Mexico)
Bill Nelson (Florida)	Bill Lowery (California)
Michael A. Andrews (Texas)	Rod Chandler (Washington)
George E. Brown, Jr. (California)	Herbert H. Bateman (Virginia)
Ralph M. Hall (Texas)	Robert S. Walker (Pennsylvania)
Mervyn M. Dymally (California)	
Norman Y. Mineta (California)	
Buddy MacKay (Florida)	
Robert G. Torricelli (New Jersey)	

*Chairman

NATURAL RESOURCES, AGRICULTURE RESEARCH AND ENVIRONMENT

Democrats	*Republicans*
James H. Scheuer* (New York)	Raymond J. McGrath (New York)
Tim Valentine (North Carolina)	Claudine Schneider (Rhode Island)
Tom Harkin (Iowa)	Rod Chandler (Washington)
Michael A. Andrews (Texas)	Tom Lewis (Florida)

*Chairman

Democrats

Buddy MacKay (Florida)
Robert G. Torricelli (New
 Jersey)
George E. Brown, Jr.
 (California)

SCIENCE, RESEARCH AND TECHNOLOGY

Democrats

Doug Walgren*
 (Pennsylvania)
George E. Brown, Jr.
 (California)
Dave McCurdy (Oklahoma)
Mervyn M. Dymally
 (California)
Norman Y. Mineta
 (California)
Buddy MacKay (Florida)
Robert G. Torricelli (New
 Jersey)
Stan Lundine (New York)
Paul Simon (Illinois)
Richard J. Durbin (Illinois)
Tim Valentine (North
 Carolina)
Harry M. Reid (Nevada)
Frederick C. (Rick) Boucher
 (Virginia)
*Chairman

Republicans

Judd Gregg (New Hampshire)
Sherwood L. Boehlert (New
 York)
F. James Sensenbrenner
 (Wisconsin)
Raymond J. McGrath (New
 York)
Joe Skeen (New Mexico)
Herbert H. Bateman
 (Virginia)

In addition to these congressmen and congresswomen, other representatives worthy of special note include Newt Gingrich of Georgia and Daniel Akaka of Hawaii. These two congressmen are co-chairmen of the Congressional Space Caucus, a group of about 120 members of the House of Representatives who are supporters of this country's space program.

Any member of Congress welcomes the opinions of the electorate, particularly when they are expressed in constructive terms (letters of complaint are the norm, not the exception; positively worded messages stand out as pleasant surprises). More than messages, however, members of Congress welcome help from people in their own district. One of the best ways to be effective in persuading a congressman or congresswoman to your cause is to give some of your time to theirs. With a little luck, an insider can even help to write the official views of the representative.

One final note: Senators are elected for six years, congressmen and congresswomen for two years, and every year there are deaths, retirements, and changes in committee composition. If you want the most recent information concerning the composition of particular committees, either call the office of your local representative and senators, or the general House of Representatives information number in Washington, 202-224-3121.

The Hardest Question

"There is more challenge in each square block of city slum than all the galaxy.
Between Brother and Brother, more awful distance,
Than the long boulevard of lonely space.
"It will be written that in 1969, primitive man canned himself
"And catapulted through the void,
"While hunger, hate and sickness stalked his earth;
"Choosing not to try for Heaven, just the Moon."
—Charles Joelson

An interest in space has carried us a long way, from the early dreams of Robert Goddard, Konstantin Tsiolkovski, and Wernher von Braun up to the moon and planets, and finally all the way back to Earth and the United States Congress.

And it is here in the center of government that the hardest questions must be answered. Space development will compete for attention and money with a hundred other programs of great urgency. The criticism will be heard again and again: How can we justify spending our time, even planning our careers, on space, when the world is full of sickness, poverty, ignorance, and suffering? When the security of our country is not assured, when crime, drugs, and violence increase daily, when our cities, churches, schools, and civil systems are falling into decay? Why should we waste money on the remote, cold, unprofitable wastes of space? Should not human needs and human suffering command all our resources and all our attention?

In Washington, in the Congress—everywhere—space supporters are constantly buffeted by these questions. Space development would be nice, they hear, but unfortunately it is a less suitable use of our limited budget than this . . . or this . . . or this. Many readers may feel disillusioned that a career search leading to the stars should terminate in the brawl of earthly politics. But unless we have sound answers to the questions of the last paragraph, we cannot with good conscience seek a career in the space program.

Fortunately, there are answers. One of the myths of Washington is that our economy works as a "zero-sum game." This term, originally introduced in the theory of games, can be explained in one short sentence: *If I win, somebody else must lose.*

In such a view of the world, there is a single pool of available resources (money, materials, manpower, and womanpower), of a certain size. Every project, from space to crime prevention to national defense to child care, draws from that same pool. If we ask more for space, goes the argument, then we must steal something from defense, welfare payments, or some other area. Your funds for the space shuttle take money from the education of my children, or prevent us from finding a cure for cancer.

The reasoning seems plausible, but the original assumption is false. The world is not a zero-sum game. *Some activities serve to enlarge the resource pool.* If a project adds more to the pool than is needed to perform it, there has been a net increase in our overall resources.

Certain activities by their nature cannot be expected to increase our resource pool. Weapons development is an excellent example. When we build a weapon—even the most necessary and unarguably essential weapon—we consume resources and provide no new resource in return. The stimulus to the economy comes from the jobs needed in building the weapons system, but it stops there. Even the research funds expended on weapons produce less than research funds spent in most other areas. By contrast, expenditure on space research tends to provide larger spin-off benefits—resource-pool growth—far beyond their costs.

Prove it, say the skeptics. If your statement is true, why is it not obvious to all? Everyone should be able to see for himself or herself when an activity amplifies the resource base, but where is the evidence that space research produces such an effect? Voyager produced pretty pictures, sure, but not real economic gains—not even Teflon frying pans or space pens. We have spent many billions on space. Where is the payoff?

Every analysis of expenditure on space shows that it *has* returned benefits, and benefits far beyond the total program costs, in many different areas. We have seen payoffs in a score of fields, from materials science, to medical telemetry devices, to electronics and computer development, to the arts. The space program has placed heavy demands on our technology and basic sciences, serving as a pacemaker and a measure of progress for dozens of different fields. And those fields have responded, again and again. And in the social areas, we've learned about humanity as we view earth from space, and as we step off the planet. But if we demand an answer that shows a return *this year* for this year's expenditure, we will not find it.

We must recognize the importance of an added factor that bears on all human affairs: time. There will always be a delay between an investment and the return on investment. And it is an unfortunate truth that the most important actions, those which will have the largest positive effect on our resources, are also likely to have the longest delays before their consequences are obvious.

If Congress and the country are looking for quick returns and an instant effect on budget deficits, they will find it easiest to cut in the area where slow returns are the rule: basic research. We will all be the losers, but who will know it this year or next year?

If we are realists, we should not look for instant returns from today's science. Half a century lay between Michael Faraday's solitary laboratory dabbling and practical electrical power, half a century between Robert Goddard's primitive rockets and the engines that launch commercial communications satellites. The payoff from the Voyager program will come in 2030, not in 1985.

So should we serve urgent present needs, leaving us with worse problems tomorrow? Or should we tighten our belts today and invest in a future that we can only dimly perceive? It is a hard decision we have to make. We can feed well now—but at our children's expense.

We are not the first generation to be faced with the unpleasant choice. Centuries ago, in times of failed harvest, there was always one desperate short-term remedy. One could eat the seed corn needed for next year's planting. Whenever this was done, it represented an absolute last resort. Families wept as they ate, knowing that today's meal meant coming starvation.

Times are hard now—they are always hard! But we are eating our seed corn when we place today's urgent social needs higher in priority than projects with better long-term

benefits. High on the list of better projects comes investment in the dizzying potential of space development.

Here then is the answer. When we seek careers in the
space program we are not shirking responsibilities or running away from problems. We are guarding the seed corn,
fighting for progress, protecting the future.

That fight will continue to be a struggle in the next decades, certainly no easier than it is now. It will take all our
energies and all our resolve. But while we work, we will
have a unique reward. Now and then we can raise our eyes
from our labors, gaze out across the span of centuries, and
discern the far-off time when today's sacrifices will be recognized as the bright spark that signaled the beginning of a
new age for humanity: the space age.

Index

AAS. *See* American Astronautical Society; American Astronomical Society

Aerospace Education Foundation, Inc., 114

Aerospace Electrical Society, 114

Aerospace Industries Association, The (AIA), 78–80

Aerospace Medical Association, 80–81

Aetna Life Insurance, 48

AGORA, 175

AIA. *See* Aerospace Industries Association

AIAA. *See* American Institute of Aeronautics and Astronautics

Air Force Association, 114

Akaka, Daniel, 229

Alouette I, 168–169

American Astronautical Society, The (AAS), 7, 81–83

American Astronomical Society, The (AAS), 84–86

American Institute of Aeronautics and Astronautics, 86–90

American Society for Aerospace Education, The (ASAE), 91

American Society of Aerospace Pilots, The (ASAP), 92–93

Ames Research Center, 41, 50, 51

AMSAT: Radio Amateur Satellite Corporation, 114

Andropov, Yuri, 200

ANIK satellites, 170

Apollo Program, 13, 33, 36, 37, 166, 211, 214

Apollo-Soyuz Test Program, 201–202, 209

APPLE satellite, 193

ARCAD III, 187

Ariane Program, 172, 173–175, 186, 191

Arianespace, 50, 175, 186

Ariel, 190

Aryabhata, 193

ASAE. *See* American Society for Aerospace Education

ASAP. *See* American Society of Aerospace Pilots

ASAT, 200, 217

Association of Lunar and Planetary Observers, 114

Astronomical League, 114

Atlas, 174

Aviation/Space Writers Association, 114

Ball Aerospace Systems, 48

Bickerton, A.W., 30–31
BIS. *See* British Interplanetary Society
Boeing, 184
Bova, Ben, 215
Braun, Wernher von, 196, 230
British Interplanetary Society, The (BIS), 93–95

California Space Institute, 114
Campaign For Space, 114
Canada Center for Space Science, 171
Canadian Center for Remote Sensing, The (CCRS), 169, 170
Canadian Department of Energy, Mines and Resources, 169
Canadian National Research Council, 172
CCRS. *See* Canadian Center for Remote Sensing
Centre National d'Etudes Spatiales. *See* C.N.E.S.
Chicago Society for Space Studies, 115
Chrétien, Jean-Loup, 187
Churchill Research Range, 171
Clarke, Arthur C., 166
Cleator, P.E., 20
CLUSTER, 176
C.N.E.S., 186–188, 189
Cobb, Geraldyn (Jerrie), 158–159
Commerce Department, 35
Communications Satellite Corporation. *See* Comsat
Comsat, 35, 47–48, 170
Conestoga I, 49
Congressional Space Caucus, 115
Congressional Staff Space Group, 115
COSPAS, 201
Council of Defense and Space Industry Associations, 115

Cousteau, Jacques, 25
Covault, Craig, 196
Cox, Dr. Anthony, 191

Defense Advisory Committee on Women in the Services, 159
Defense Department, 44–45, 46, 47, 50, 133–134, 139, 158–159. *See also* Space Command
Delta, 174
Delta Vee, Inc., 95–96
Department of Aerospace Engineering, 114
Department of Agriculture, 46
Department of Energy, 211–212
Dryden Flight Research Center, 42, 50

Earth Resources Satellite. *See* ERS-1
Earth Resources Technology Satellite. *See* ERTS
Earthnet system, 177
EISCAT, 185
ERS-1, 177, 180–181, 191, 210
ERTS, 169, 210
ESA. *See* European Space Agency
ESRANGE, 185
EURECA, 173
European Communications Satellites (ECS), 177, 191
European Incoherent Scatter Scientific Association. *See* EISCAT
European Retrievable Carrier. *See* EURECA
European Space Agency, The (ESA), 172–177, 186, 191, 210
EXOSAT, 175

Fairchild Space and Electronics, 48

Federal Women's Program, 163–164
FIRST, 176
Ford Aerospace, 179
Foundation for Scientific Progress And Continual Exploration (SPACE), 115
Fox, Joseph, 58–59
Fullerton, Gordon, 195

Gagarin, Yuri, 166
Galileo project, 24, 127, 187, 212
Gates, Anita, 128
G.D.T.A., 189
Gemini programs, 158
General Electric, 179
GEOSAT Committee, The, 115
Gingrich, Newt, 229
Giotto, 175
Glaser, Peter, 211
Glenn, John, 158–159, 166, 223–224
Goddard, Robert, 29–30, 31, 196, 230
Goddard Institute for Space Studies, 44
Goddard Space Flight Center, 41–42, 50, 51
Groupement pour le Développement de la Télédétection Aérospatiales. See G.D.T.A.

Hakucho (Swan) satellite, 182
Haldane, J.B.S., 20
Hermes, 174, 181
Herschel, William, 208
Herzberg Institute of Astrophysics, 171
Hewlett-Packard, 75
High Frontier, Inc., The, 115
Himawari program, 179, 180

Hipparcos, 175
Honitori (Firebird) satellite, 182
Huey, Mary Evelyn Blagg, 159
Hughes Aircraft, 47, 179
Hypatia Cluster, The, 96–97, 164

IAF. See International Astronautical Federation
IBM, 48, 119, 125
IEEE. See Institute of Electrical and Electronics Engineers
Independent Space Research Group, 115
Indian Space Research Organization (ISRO), 192–193
INSAT, 193–194
Institute for Security and Cooperation in Outer Space, 115
Institute for Space Research, 201
Institute for the Social Science Study of Space, 115
Institute of Electrical and Electronics Engineers, The (IEEE), 98–100
Institute of Space and Astronautical Science (ISAS), 178, 181–183
Institute of Space and Security Studies, 115
Intelsat, 170
International Alliance for Cooperation in Space, 115
International Astronautical Federation, The (IAF), 101–102
International Solar Polar Mission, 175
International Telecommunications Satellite Organization. See Intelsat
IRS, 193
ISAS. See Institute of Space and Astronautical Science

ISRO. *See* Indian Space Research Organization
Itokawa, Hideo, 178

Jet Propulsion Laboratory, The, 43, 51
Joelson, Charles, 230
Johnson, Lyndon Baines, 216
Johnson, Nicholas, 196
Johnson Space Center, 26, 43, 44, 50, 51

Kagoshima Space Center, 178
Kennedy Space Center, 26, 42, 44, 50, 51
Kepler, Johannes, 24
Khrushchev, Nikita, 157
Korolev, Sergey Pavlovich, 196
Kramer, Saunders, 196

LAMBDA 4-S, 178
Landsat satellites, 169, 193, 194, 210
Langley Research Center, 42, 50, 51
Large Communications Satellite (L-SAT), 177, 191
Lewis Research Center, 43, 51
L-5 Society, The, 7, 102–103
L.M. Ericsson, 184
LOS satellites, 169
Lousma, Jack, 195
Luna program, 21, 165
Lunar Orbiter, 165–166

Macdonald Dettwiler & Associates, 170
McDonnell Douglas Astronautics, 48–49, 182
MARECS, 177, 191
Marine Observation Satellite (MOS-1), 180
Mariner, 166

Maritime Communications Satellite. *See* MARECS
Marshall Space Flight Center, 42, 43-44, 50, 51
Mars program, 165
Maryland Space Futures Association, 115
Mercury Program, 157–158, 159, 195
Meteosat satellites, 177, 191
Michoud Assembly Facility, 43–44
Ministry of Defence, Great Britain, 190
Ministry of Information Technology, Great Britain, 190
Mitsubishi Electric Corporation, 179, 182
Mobile Satellite Program (MSAT), 170–171
Mobile Scientific Balloon Launch Facility, 171
Moore, Patrick, 20–21
MOS satellites, 169
MSAT. *See* Mobile Satellite Program

NASA, 19, 22–23, 31–51, 60–61, 133–134, 139, 157–158, 163–164, 167, 171, 173, 178, 179–181, 182, 187, 196, 209, 212–213, 215–216, 219–220
NASDA, 178, 179–181, 182
National Advisory Council for Aeronautics (NACA), 32
National Aeronautics and Space Administration. *See* NASA
National Aeronautics Establishment, 171
National Environmental Research Council (NERC), 190
National Oceanic and Atmospheric Administration (NOAA), 35

National Remote Sensing Programme Board, 190
National Space Club, The, 103–105
National Space Development Agency of Japan. *See* NASDA
National Space Institute, The (NSI), 7, 105–107
National Space Technology Laboratories, The, 43
New York Times, 29–31
Niagara University Space Settlement Studies Project, 116
Nippon Electric Company, 179, 182
Nissan Motors, 182
Nordsat direct broadcast satellite, 183
NSI. *See* National Space Institute

OASIS (Organization for the Advancement of Space Industrialization and Settlement), 116
Oberg, James, 196
Oberth, Hermann, 29, 30, 196
Office of Management and Budget (OMB), 36, 61, 178, 219–220
OHSUMI, 178
Oreol 3 satellite, 187
OTS-1 test satellite, 176–177

Pioneer, 166, 209
Planet-A Halley's Comet probe, 182
Planetary Society, The, 107–108
Progressive Space Forum, 116
Proxmire, William, 218
Prudential Insurance Company, 49

Ranger, 165

RCA Astro Electronics, 35, 47, 179
Rohini satellite, 192
Royal Aircraft Establishment, 190
Royal Astronomical Society of Canada, 116
Ryumin, Valery, 209

Saab-Scania, 184
Salyut, 48, 187, 197–198, 209
San Francisco Space Frontier Society, 116
SARSAT, 201
Satellite Business Systems, 48
Satish Dhawsan, 193
Science and Engineering Research Council (SERC), 190
Seasat, 169
SEDS. *See* Students for the Exploration and Development of Space
Senate Subcommittee on Science, Technology and Space, 224
Sheldon, Charles, 196
Shepard, Alan, 166
Skylab, 37, 166, 209
Skynet, 190, 191
SLV-3, 192
SOHO, 176
Solar Electric Propulsion System (SEPS), 212
Soyuz, 197
Space Act of 1958, 32
Space Cadets of America, 116
Space Coalition, The, 116
Space Command, 143–146
Space Development Foundation, 116
Space Foundation, The, 116
Space Industries, Inc., 48
Space Services, Inc., of America (SSI), 49
Space Studies Institute (SSI), 108–109

Space Telescope, 187
Space Transportation Company, Inc. *See* SpaceTran
SPACECOM. *See* Space Command
Spacelab, 171, 172–173, 176, 182, 187, 201
SpaceTran, 49
Spaceweek, Inc., 109–111, 116
Spar Aerospace, 168
SPOT IMAGE, 189
SPOT satellites, 169, 177, 188–189, 193, 210
Sputnik 1, 32, 165, 195
SSC. *See* Swedish Space Corporation
SSI. *See* Space Studies Institute
Starlab, 171
Students for the Exploration and Development of Space (SEDS), 111
Sunsat Energy Council, 116, 212
Surveyor, 165
Svenska Rymdbolaget, 184
Swedish Space Corporation (SSC), 184
Système Probatoire d'Observation de la Terre. *See* SPOT

Tanega-shima Space Center, 180
Technology Laboratories, 51
Telesat Canada, 170
Tele-X, 183–184
Tereshkova, Valentina, 157, 166
Texas Instruments, 75
Texas Woman's University, 159
TEXUS, 176
Titan, 174
Toshiba Corporation, 179, 182
Tsiolkovski, Konstantin, 29, 30, 196, 230
Tsukuba Space Center, 179–180

United Futurist Association, 116

United States, departments of. *See under specific departments, e.g., Defense Department*
United States Space Education Association, 116
Universities Space Research Association (USRA), 64, 111–113
University of California Space Working Group, 116
Upper Atmosphere Research Satellite (UAR), 187
Using Space for America (USA) Committee, 117
USRA. *See* Universities Space Research Association

VEGA mission, 187
Venus Orbiting Imaging Radar (VOIR) mission, 209–210, 212
Viking satellite, 166, 172, 184–185
VOIR, 212
von Puttkamer, Jesco, 23
Voyager, 127, 166
V-2 rocket, 32

Wallops Island Flight Center, 43, 51
War Control Planners, Inc., 117
Webster, Daniel, 218
Western Union, 47
White Sands Test Facility, 44
World Security Council, 117
World Space Center, 117
World Space Foundation, 117
Wright brothers, 28
WRITE NOW!, 117

XMM, 176
X-ray Astrophysics Facility (AXAF), 212

Zaitsev, Yuri, 201–202